From Promises to Performance

From Promises to Performance
ACHIEVING GLOBAL ENVIRONMENTAL GOALS

Gary C. Bryner

BRIGHAM YOUNG UNIVERSITY

W. W. NORTON & COMPANY

NEW YORK / LONDON

Copyright © 1997 by Gary C. Bryner

All rights reserved

Printed in the United States of America

The text of this book is composed in Janson
with display set in Onyx Compressed and Bodoni
Composition by Special Projects Group
Manufacturing by Haddon Craftsmen, Inc.
Book design by Jam Design adapted by Jack Meserole

Library of Congress Cataloging-in-Publication Data
Bryner, Gary, 1951–
From promises to performance: achieving global
environmental goals /
Gary Bryner
 p.   cm.
Includes index.
1. United States—Foreign relations—Environment
2. Environment—Foreign relations—United States
I. Title.
E183.8.I57J46   1997
327.730567—dc20   97-1610

ISBN 0-393-97172-4

W. W. Norton & Company, Inc., 500 Fifth Avenue, New York, N.Y. 10110
http://www.wwnorton.com

W. W. Norton & Company Ltd., 10 Coptic Street, London WC1A 1PU

1 2 3 4 5 6 7 8 9 0

To my parents, for all they have taught me

Contents

	Acknowledgments	ix
	Introduction	1
1	International Environmental Agreements in Place	14
2	Challenges in Achieving the Goals of Global Environmental Agreements	87
3	Alternatives for Achieving International Environmental Goals	133
4	International Institutions for Protecting the Environment	173
5	The Political Economy of Global Environmental Regulation	215
6	Environmental Preservation and the Less-Developed World	257
7	Prospects for Preserving the Global Environment	305
	Index	345

Acknowledgments

I BEGAN WRITING this book five years ago as an attempt to try to understand what is involved in accomplishing the ambitious goals reflected in global environmental agreements and how we might move toward a more ecologically sustainable world. Along the way I have become encouraged by some modest glimmerings of progress and sobered by the tremendous challenges that remain. I very much hope that the book will be useful to students and others who, like me, are struggling to understand what we need to do to remedy the global environmental problems that pose such profound risks to current and future generations. I am also most grateful for the opportunity to acknowledge the help I have received from many people and institutions.

I teach a two-semester course with Professor Samuel Rushforth of the Botany Department at Brigham Young University on ecological science and environmental policy, and that course has been enormously important to me in learning about global ecology. At the end of the winter semester, we have traveled as a class to Latin America to see firsthand some of the challenges involved in achieving global environmental and development goals. We have also worked alongside villagers on projects they have designed to make their communities a little more ecologically sustainable and to improve their quality of life. Several of these trips have been organized through the efforts of the Center for Humanitarian Outreach and Intercultural Exchange (CHOICE) and the Andean Children's Foundation; others have been organized through the Tarahumara Foundation. I am greatly indebted to Suzanne Lundquist, Scott Abbott, Timothy Evans, Neils Valentiner, Jaime Figueroa, Ted Lyon, Eran Call, Carolyn Dailey, James Mayfield, Humberto Ramos, and many others in Bolivia, Mexico, and Peru whose names I do not know.

I have presented versions of some of the chapters in this book at conferences organized by the American Political Science Association, the Association for Policy Analysis and Management, the International Studies Association, and the Western Political Science Association. I am most grateful to colleagues who have offered comments and criticisms and included me on panels where I have learned much from other scholars. Panels organized by the Environmental Studies Section of the International Studies Association have been particularly helpful as forums for discussion about the issues I address here. I am greatly indebted to scholars in political science, economics, law, environmental science, and other disciplines; references to their work in the endnotes of this book do not begin to acknowledge how much I have learned from them.

A number of talented research assistants at BYU have made major contributions to this project: Barry Balleck, Dylan George, Emily Lauritzen, Tyler Rushforth, and Margaret Wooley. The David M. Kennedy Center for International Studies and the College of Family, Home, and Social Science at BYU have provided generous assistance in support of my research. Courses I took at the BYU and Vermont law schools were an essential part of my preparation for the research in this book, and I am most grateful for the support I received from BYU and from Stan Albrecht, Clayne Pope, Dennis Thomson, Stan Taylor, and David Magleby in particular in helping to provide that opportunity.

Acknowledgments

I have had the great fortune to come to know Roby Harrington and Steve Dunn and the editors and production staff at W. W. Norton, who have provided great support for and assistance with this book.

Finally, my family—Jane, Ben, Nick, and Jonathan—have been of great help and support in many ways as I have worked on this project. My parents, Barbara and Clifford Bryner, have taught me much during my life. It has been particularly pleasant to reflect on all I have learned from them and to have the opportunity to dedicate this book to them.

Introduction

THE SIERRA NEVADA de Santa Marta in Colombia is the highest coastal mountain range in the world, rising some 5,800 meters from sea level. It is home to the Kogi, people who call themselves the "elder brothers" of humankind. They are among the last functioning pre-Columbian societies in South America. Kogi families regularly move among homesteads, rotate their crops from field to field, and walk from one valley to the next in sacred patterns of movement. They have an intimate knowledge of their land, a knowledge they consider to be a sacred legacy from the Ancient Ones. The Kogi carefully cultivate knowledge of the workings of the universe, and they believe they can communicate with other spheres of existence.[1]

The Kogi believe they must use their special knowledge to find a balance between the creative and destructive energies of life. In 1990, they ended their isolation from the rest of the world's people—whom they call "younger brothers"—to issue a warning concerning the damage humans are doing to the earth and to other living things. The Kogi believe that ignorance and greed will bring an end to life on earth. The Kogi chose the British historian Alan Ereira to communicate their message that humankind has failed to think,

INTRODUCTION

behave, and "feel the world" in ways that ensure that life as we know it can continue.

"How is it that we are able to live?" they ask.

It is the mountains which make the waters, the rivers, and the clouds. If their trees are felled they will not produce any more water. We do not cut down the trees that grow by rivers, we know that they protect the water. We do not cut down huge areas of forests like the Younger Brother does, we cut small clearings for our fields. The Mother told us not to cut down many trees, so we cut few, tiny patches. If the Younger Brothers keep cutting down all the trees, there will be fires because the sun will heat the earth.

Younger Brother, stop doing it. You have already taken so much. We need water to live. Without water we die of thirst. We need water to live. The Mother told us how to live properly and how to think well. We're still here and we haven't forgotten anything.

The earth is decaying, it is losing its strength because they have taken away much petrol, coal, many minerals. Younger Brother thinks, "Yes! Here I am! I know much about the universe!" But this knowing is learning to destroy the world, to destroy everything, all humanity. . . .

Does the Younger Brother understand what he has done? Does he?[2]

The Kogi are not alone in their concern about our environmental predicaments. The Worldwatch Institute, established in 1974, serves as an early warning system for the global environment. Its recent reviews describe the state of the planet in somber terms. The stratospheric ozone layer is thinning even faster than was thought just a few years ago; scientists have estimated that for every 1 percent decline in atmospheric ozone, there is a 2 percent increase in the amount of ultraviolet radiation that reaches the earth's surface. At least 140 plant and animal species, on average, become extinct each day. Atmospheric levels of heat-trapping carbon dioxide and other greenhouse-effect gases are much higher than preindustrial concentrations, and continue to climb. The earth's surface was warmer in 1990 than in any year since record keeping began in the mid-nineteenth century; six of the seven warmest years on record have occurred since 1980.[3] The world's forests are disappearing at a rate of some 17 million hectares per year. Global population is growing by 92 million people a year, a number equal to the population of Mexico; the vast majority of the increase, 88 million, takes place in the developing world.[4] Some 26 billion tons of topsoil are lost each year beyond the formation of new soil, and 6 million hectares of new desert are formed each year because of poor land management. Thousands of lakes in the developed nations are biologically dead, and thousands more are dying. Water tables are falling throughout the world as demand overwhelms aquifer recharge rates.[5]

The doubling of the world's population, the quintupling of global economic output, and the widening gap in the distribution of income that have occurred during the past four decades have resulted in a world where 800 million people are malnourished and another billion are undernourished, 1.3 billion live in absolute poverty, 34,000 children die each day from malnutrition and disease, and 35 million people are refugees from violence. In many parts of the world, we have exceeded the capacity of natural resources and social systems to provide a decent standard of living.[6] The depletion of natural resources has already resulted in adverse economic effects such as lost jobs, reduced exports, and industry shutdowns. Many countries with high population growth rates face large deficits in food production in the future. Although there is great disagreement concerning how many people the earth can support, three limits in global natural resources appear to be responsible for slowing the growth of world food production: (1) the sustainable yield of oceanic fisheries; (2) the amount of fresh water produced by the hydrological cycle; and (3) the amount of fertilizer that existing crops can effectively use.[7]

In 1993, some 1,670 scientists from seventy-one countries, including the majority of living Nobel Prize winners, signed a "Warning to Humanity" that argued human beings and the natural world "are on a collision course." Current practices "cannot be continued without the risk that vital global systems will be damaged beyond repair." We have little time to respond: "no more than one or a few decades remain before the chance to avert the threats we now confront will be lost and the prospects for humanity immeasurably diminished." Some problems are global in nature, while others are primarily local problems found throughout the world. The "warning to Humanity" focused on six critical environmental issues:

The Atmosphere. Stratospheric ozone depletion threatens us with enhanced ultraviolet radiation at the earth's surface, which can be damaging or lethal to many life forms. Air pollution near ground level and acid precipitation are already causing widespread injury to humans, forests, and crops.

Water Resources. Heedless exploitation of depletable groundwater supplies endangers food production and other essential human systems. Heavy demands on the world's surface waters have resulted in serious shortages in some 80 countries containing 40 percent of the world's population. Pollution of rivers, lakes, and groundwater further limits the supply.

Oceans. Destructive pressure on the oceans is severe, particularly in the coastal regions, which produce most of the world's food fish. The total marine catch is now at or above the estimated maximum sustainable yield. Some fisheries have already shown signs of collapse. Rivers carrying heavy burdens of eroded soil into the seas also carry industrial, municipal, agricultural, and livestock waste—some of it toxic.

Soil. Loss of soil productivity, which is causing extensive land abandonment, is a widespread by-product of current practices in agriculture and animal husbandry. Since 1945, 11 percent of the earth's vegetated surface has been degraded—an area larger than India and China combined—and per capita food production in many parts of the world is decreasing.

Forests. Tropical rain forests, as well as tropical and temperate dry forests, are being destroyed rapidly. At present rates, some critical forest types will be gone in a few years, and most of the

Introduction

tropical rain forest will be gone before the end of the next century. With them will go large numbers of plant and animal species.

Living Species. The irreversible loss of species, which by 2100 may reach one-third of all species now living, is especially serious. We are losing the potential they hold for providing medicinal and other benefits, and the contribution that genetic diversity gives to the robustness of the world's biological system and to the beauty of the earth itself.

Underlying all these environmental problems is the earth's burgeoning population:

> Pressures resulting from unrestrained population growth put demands on the natural world that can overwhelm any efforts to achieve a sustainable future. If we are to halt the destruction of our environment, we must accept limits to that growth.[8]

The nations of the earth have signed more than 175 international environmental agreements on subjects ranging from fishing to conservation of plants to limiting ocean pollution. The 1987 Montréal Protocol on Substances that Deplete the Ozone Layer, the climate change and biodiversity treaties signed at the 1992 Earth Summit in Brazil, the Law of the Sea treaty, and a number of other accords are significant milestones in global cooperation and historic achievements of global environmental diplomacy. They have created tremendous expectations and demands for effective implementation. But they represent only the beginning of a long series of decisions and actions that confront nations. Although many environmental problems and threats are global in scope and in consequence, the solutions ultimately lie in changing individual behavior. Heads of state and official delegations may debate issues and haggle over the language of agreements, but much of the progress will have to come from the efforts of grass-roots organizations, government bureaucrats, local communities, families, and individuals who are committed to changing the way they and others interact with their environment.

INTRODUCTION

A tremendous scientific debate rages over global warming, deforestation, extinction of species, ocean pollution, acid rain, loss of topsoil, and a host of other problems: just as challenging is the policy debate over how to bring about the political, economic, and social changes that will be necessary to prevent or mitigate their effects. The difficulties in negotiating and gaining support for international agreements are formidable: just as perplexing is the question of how to ensure that once agreements are in place, they are effectively carried out and their goals are achieved. The challenges involved in translating international agreements, statements of principles, and general expressions of concern into concrete actions by individuals, families, corporations, local communities, and governments are daunting.

The preservation of natural resources and ecosystems and the reduction of environmental pollution pose profound challenges for all societies. These problems are particularly difficult to address in democracies, since they are infused with uncertainty, complex interactions, and long-run consequences that threaten to overwhelm traditional processes and institutions. Philosophical inquiry and analysis, studies of the political process, and assessments of specific policies are all part of the enterprise.[9] Given the political context in which these decisions must take place, theories and concepts from political science help illuminate the debate. Given the importance of these issues, the way in which we collectively address them has serious implications for our understanding of the nature of politics and government.

A number of broad questions outline the issues at the heart of global environmental threats. How can a society avoid fundamental, sweeping changes with little discussion or deliberation? Given the importance of trial and error in political decision making, how can we enhance our capacity to learn from such experiences? How can public discussion occur early in the process of technological change, so that it takes place soon enough to shape alternatives and explore possibilities, rather than later when interests are cemented, resources are invested, and adjustments are difficult? How can we slow the pace of innovation so that we can monitor, interpret, debate, and revise the original developments? What modifications

Introduction

of institutional structures, processes, and elite incentives might allow more reliable and more timely perception, interpretation, and utilization of feedback? How can we provide for mutual accommodation among diverse partisan interests and ensure that technological decisions are responsive to ordinary people's concerns and needs? Given the political nature of technologies, how can we make business decisions more responsive to consumers and workers? How can we enhance the capacity of government and business institutions so that scientific, technological, and environmental considerations are more effectively addressed? How can we foster institutional flexibility?[10]

One group of advocates may hold that we are in the midst of an environmental/technological crisis that threatens the survival of the human race. Nothing short of a fundamental revolution in our patterns of production and consumption will suffice; we need to transform our culture and way of thinking. Another group may believe we need devise only modest incentives that will produce a moderate shift in our behavior. Since public policy making is largely an incremental process, is radical change even possible? The uncertainties surrounding many environmental problems may make it difficult to decide on appropriate comprehensive actions. Incremental policy making may permit us to learn as we go along and make adjustments in policies as new information is received. Incremental changes may be inadequate, however, given the nature of the environmental threats that confront the world.

My purpose in writing this book is to join the debate over international environmental problems and their solutions by focusing attention on the goals underlying global environmental agreements and the policy choices we face in pursuing those goals. I am particularly interested in exploring ways we might answer six questions that are central to the challenge of preserving the global environment:

> What actions are required to implement global environmental accords and achieve their goals?

> What are the most promising public policy tools for accomplishing these goals more efficiently and effectively?

INTRODUCTION

How can we build the institutional capacity required to ensure that global environmental goals are achieved?

How can we integrate the efforts called for by global accords with the goals of economic growth among the more developed countries?

How can we pursue environmental protection and remedy the problems facing the less developed nations?

Finally, even if we achieve the goals of global environmental agreements, will that be sufficient to preserve the planet?

Although the primary focus of this book is on policy choices confronting the United States, the global character of these challenges means that other nations face these choices as well.

I hope readers of this book will gain a clearer sense and appreciation of the choices we face in sustaining the global biosphere for the rest of our lives and for those who come after us. I particularly hope it will be useful to students of politics and the public policy-making process as an illustration of the challenges in making policy, to students of environmental science as an overview of some of the political and policy issues involved in responding to environmental problems, to students of international relations who are studying international environmental law and policy, and to others who have a general interest in preserving the global environment. The solutions to the problems discussed here require broad political action that can only arise from a wide-ranging discussion of the issues, choices, and policy options. The kinds of changes to be made are overwhelming in their scope, complexity, and cost. Broad acceptance of the need for action and a willingness to make these changes are essential.

For those who believe there is no compelling evidence that global environmental agreements are needed, this discussion may appear premature. In spite of great uncertainties and disagreements among scientists, however, the magnitude of the possible consequences and how they are distributed throughout the world require us to take preventative or precautionary action. While the debate over

Introduction

the likelihood and magnitude of global environmental changes continues, we need to explore now how to bring about the changes that will be needed.

The United States and other wealthy countries have a particular *obligation* to comply with these international agreements. Americans pollute more and consume more resources than any other people. The United States is so economically and politically powerful that its participation in global environmental protection efforts is essential. Americans have an obligation to help reduce the risks that confront all humanity, especially people in the less developed world who do not have the resources to escape, at least in the short term, many of the consequences of global environmental problems. Mitigation of the effects of stratospheric ozone layer depletion or global climate change, for example, will much more likely occur in the wealthy world, among the people who are primarily responsible for these threats, than among the poor of the world. The poor, who have not benefited as much from the industrial and commercial activities that are responsible for environmental threats, largely bear the environmental risks of these activities. It is also in our self-interest to make global agreements work, since ultimately we cannot escape many of the consequences of environmental degradation. The stakes are extremely high: if we do not prevent the environmental calamities that are forecast, it does not really matter what else we try to do. The planetwide consequences may simply be irreversible.

These obligations also represent *opportunities* to address a number of important global concerns. We can protect natural resources and environmental quality and at the same time improve the prospects for those in the less developed world. We can achieve multiple goals, many of which are clearly noncontroversial and certain. Reducing emissions of greenhouse gases will reduce energy consumption and thereby increase national security, reduce energy costs and save money, reduce local pollutants that pose health threats, and produce other benefits. Certain powerful interests seek to preserve the status quo and the benefits they derive from our current ways of thinking and doing, but if we are willing to challenge these interests we will find in many areas that the long-term

INTRODUCTION

benefits to all greatly exceed the short-term disruptions and losses. We can help ensure that the pain these transitions impose is fairly distributed, but we cannot escape difficult choices and the need to generate strong public support to make those choices.

We are at an extraordinary moment in history. Decisions made during the next few decades may determine the very survivability of humans, as well as many other species with which we share the planet. There are in this debate, like most complex debates, many more questions than answers. The objective of this book is to help frame those questions in ways that will contribute to public understanding, discussion, and eventually, effective responses.

Outline of the Book

Chapter 1 reviews the provisions of current international environmental agreements and the commitments the United States and other nations have already made, and outlines the kinds of programs nations must undertake to achieve global environmental goals. Even a brief discussion of all the international environmental accords in place is beyond the scope of this chapter. The discussion focuses on the major agreements that seek to protect the atmosphere, covering such issues as acid rain, the ozone layer, and climate change; marine resources and Antarctica; biodiversity and endangered and threatened species; and checking population growth. Although not exhaustive, these agreements treat some of the most important obligations and opportunities the United States and other nations face.

Chapter 2 examines the difficulties inherent in implementing public policies and focuses on the particular challenges posed by global environmental agreements and the kinds of actions required to accomplish their goals. Students of the policy-making process have developed a number of models for assessing policy implementation that have been useful in designing a conceptual framework for global environmental agreements. The chapter centers on the nature of environmental problems and the challenges they pose in

Outline of the Book

devising effective policy responses; the broad political, economic, and cultural context in which these policies are developed and implemented; the legal and other policy instruments and devices available to bring about changes in global behavior; the interaction of incremental and comprehensive policy efforts; and the institutions that are most likely to contribute to effective policy responses. The primary question asked here is, What is required for effective implementation of global agreements to take place? The discussion here does not focus on the implementation of specific agreements and treaties, but provides an overview of the issues.

Chapter 3 continues the discussion of implementation and focuses on the debate over making environmental regulation more effective by increasing the use of market-based regulatory instruments. Since the list of possible regulatory interventions overwhelms the resources available, many policy makers advocate minimizing the costs of accomplishing environmental goals so that more actions can be undertaken. Questions here include, How can we accomplish environmental goals as efficiently as possible? and What kinds of legal instruments are available to remedy global environmental problems and implement global agreements?

Chapter 4 broadens the discussion to focus on the kinds of international institutions that are needed to ensure effective implementation of global agreements. We face a tremendous range of policy choices that will require a complex matrix of individual, local, regional, national, and international actions. Many countries lack the regulatory infrastructure—laws, regulations, research capabilities, and enforcement mechanisms—to support new regulatory programs for implementing global agreements.? Are the United Nations and other existing institutions capable of ensuring effective implementation of global agreements, or are new institutions needed?

Chapter 5 broadens the discussion further by exploring the global context in which environmental policies are formulated and their interaction with other public concerns, particularly economic growth. Some level of environmental quality is ultimately a precondition for economic activity: How can short-term economic imperatives be balanced with long-term preservation needs? How can environmental preservation efforts be pursued alongside economic competitiveness, trade, growth, and other policy goals that play a central role in the more developed countries of the world?

Chapter 6 focuses on the less-developed world. The residents of these nations who are beset by basic economic, social, and political problems may view those demands as more urgent than environmental ones. They are plagued with old, highly polluting technologies but have few resources to modernize their industrial and energy-producing sources of pollution. They have fewer resources to make needed changes in industrial, agricultural, and conservation activities than their wealthier neighbors. They also have more to lose than do industrialized nations, since they have few resources to mitigate the effects of ozone depletion, water pollution, or climate change. How can we balance environmental preservation with efforts to improve the material quality of life in less developed countries? How should the interests of future generations be considered in making collective decisions, and how do those concerns interact with our commitments to current generations, particularly those in the less developed world, to improve their quality of life?

The final chapter focuses on the uncertainties surrounding the causes and consequences of global environmental problems. Even if we are successful in responding to the challenges outlined in the previous chapters, will that be enough? Are the global agreements in place sufficient to ensure that the current generation can meet its material needs, and to preserve the planet for future generations? Are incremental policy efforts enough, or are more fundamental changes required? Are fundamental changes in economy and society likely to occur only in response to a crisis, or can we devise plans to move to a more environmentally sustainable political economy?

References

1. T. C. McLuhan, *The Way of the Earth* (New York: Simon & Schuster, 1994), 362–64.
2. Alan Ereira, *The Elder Brothers* (New York: Vintage Books, 1990), 163, 217, 224–25.
3. Lester R. Brown, "Nature's Limits," in *State of the World 1995*, ed. Lester R. Brown et al. (Washington, D.C.: Worldwatch Institute, 1995), 1–20, at 4.
4. Sandra Postel, "Denial in the Decisive Decade," in *State of the World 1992* ed. Lester Brown et al. (New York: Norton, 1992), 3.
5. Worldwatch Institute, reported in "The State of the World: An interview with Lester Brown," *Technology Review* (July 1988): 51–58.
6. Sandra Postel, "Denial in the Decisive Decade," in *State of the World 1992*, ed. Lester R. Brown et al. (New York: Norton, 1992), 3.
7. Brown, "Nature's Limits," at 3.
8. "World Scientists' Warning to Humanity," distributed by the Union of Concerned Scientists (April 1993).
9. For a helpful overview of these issues, see Patrick W. Hamlett, *Understanding Technological Politics* (Englewood Cliffs, N.J.: Prentice-Hall, 1992).
10. For more on these issues, see Joseph G. Morone and Edward J. Woodhouse, *Averting Catastrophe: Strategies for Regulating Risky Technologies* (New Haven, Conn.: Yale University Press, 1986); Sheila Jasanoff, *The Fifth Branch: Science Advisers as Policymakers* (Cambridge, Mass.: Harvard University Press, 1990); Charles E. Lindblom and Edward J. Woodhouse, *The Policy-Making Process* (Englewood Cliffs, N.J.: Prentice Hall, 1993); H. W. Lewis, *Technological Risk* (New York: Norton, 1990); and Charles Perrow, *Normal Accidents: Living with High-Risk Technologies* (New York: Basic Books, 1984).

1

International Environmental Agreements in Place

IN 1972 leaders from several nations met in Stockholm for the first United Nations Conference on the Human Environment. That eleven day conference formally launched a global concern with the health of the planet, a concern that has grown each year. The Stockholm meeting, preceded by two years of preparatory meetings, research, and planning, signaled a new consciousness of the earth's ecology and reflected optimism about the ability of humans to protect their planet. All member countries of the United Nations were invited to participate in the conference and in the preparation of reports on the state of their own environment. More than one hundred thousand pages of preliminary documents were prepared for the meeting, and 40 tons of documents were circulated during the sessions.[1]

Not all nations endorsed the Stockholm earth summit. Communist-bloc countries derided concern over pollution: it was a capitalist problem from which they did not suffer. Many residents of the developing world regarded poverty was a much greater and more immediate threat, and they feared that environmentalism would limit their efforts to reduce hunger and suffering. Environmentalism was a luxury they could not afford, and a potential barrier to economic development, perhaps even a plot by the wealthy nations to subjugate them.[2] But for other countries, the Stockholm conference took place in a heady atmosphere where participants could assert that "the capability of man to improve the environment increases with each passing day."[3]

Maurice Strong, the secretary-general of the conference, commissioned Barbara Ward and René Dubos to provide a conceptual framework for the delegates to read in preparation for the conference.[4] Their report, *Only One Earth*, argued for a collective international responsibility for understanding the human environment, the creation of global policies to address international concerns, and a recognition of the interdependency of life on earth and the importance of protecting the environment.[5] The Stockholm Declaration, the culmination of the meeting, was a brief statement of twenty-six "common principles to inspire and guide the peoples of the world in the preservation and enhancement of the human environment." These principles emphasized the importance of (1) safeguarding natural resources and wildlife for present and future generations, (2) preventing pollution that could not be safely absorbed by the environment, (3) accelerating economic and social development in the poorer nations, (4) encouraging the free flow of scientific information, and (5) eliminating nuclear weapons and the threat of mass destruction.[6] Two important outcomes of the Stockholm conference were (1) a commitment to collect data on environmental trends and problems and (2) the creation of the United Nations Environment Program (UNEP).

Although economic, political, and national security interests largely shaped and dominated national and international efforts, the UNEP and other UN organizations, along with a host of nongovernmental organizations (NGOs), made modest inroads in calling attention to environmental problems throughout the 1970s and 1980s. The conference spawned a series of meetings on population, food and agriculture, women's rights, desertification, human settlements, science and technology, and other issues. In the 1970s, a number of treaties, conventions, and agreements to limit ocean dumping and protect endangered species were added to earlier treaties protecting Antarctica and banning atmospheric nuclear tests. A major step in global cooperation occurred in 1987 with the signing of the Montréal agreement to phase out chemicals responsible for ozone layer depletion.

In retrospect, the list of environmental problems on the Stockholm agenda was relatively modest: the growing presence of pesticides and other toxic chemicals in the biosphere, declining whale population, ocean pollution, radioactive fallout from atmospheric nuclear tests, loss of topsoil, and an exploding global population, especially in the developing world.

Twenty years after Stockholm, the United Nations Conference on the Environment and Development (UNCED), dubbed the "Earth Summit," met in Rio de Janeiro. This time, the agenda was much longer, more complicated, and even more sobering. The 1992 meeting was an extraordinarily ambitious effort to consider the plight of the planet's ecosystems, and although it fell far short of the organizers' hopes, the delegates debated and ratified a number of agreements. The Rio Declaration, or "Earth Charter," which the delegates adopted by consensus, is a brief statement calling the world to action and outlining the broad responsibilities of the rich and poor nations. It recognizes a "right to development" for the poorer nations and calls on all nations to commit to assisting with the "developmental and environmental needs of present and future generations."[7]

Although the Rio Declaration itself is nonbinding and was dismissed by some as meaningless political posturing, others hailed its significance: leaders of most of the nations of the earth had signed an agreement that represents at least some commitment to sustainable development and environmental preservation.[8] The declaration reaffirmed that states have "the sovereign right to exploit their own resources pursuant to their own environmental and development policies," although they also have the "responsibility to ensure that activities within their jurisdiction or control do not cause damage to the environment of other States or of areas beyond the limits of national jurisdiction."[9]

The summiteers also reviewed and adopted by consensus Agenda 21, an eight-hundred page packet of twenty-nine documents out-

International Environmental Agreements

lining what nations and international organizations agreed to do to protect the environment and promote sustainable development in the Third World. The various chapters called on participating states to make a number of commitments, including those to

- develop ways to achieve sustainable levels of consumption in the industrialized nations;

- address population growth "where appropriate";

- consider market-oriented reform of their economies;

- encourage prices to be set that incorporate and internalize the environmental costs of production and disposal;

- ensure increased participation by women in development and environmental policies and programs;

- support the creation of a new agency, the UN Commission on Sustainable Development, to collect data on environmental and development activities and to monitor implementation of the provisions of Agenda 21 by participating states through "national action plans";

- facilitate the transfer of technologies from the developed to the developing world;

- take actions to maintain or increase biodiversity;

- eliminate subsidies for harvesting of natural resources that "do not conform with sustainable development objectives";

- expand their institutional capacity to encourage sustainable development;

- increase access to natural resources by indigenous peoples and expand agricultural training and assistance in rural areas;

- strengthen governmental capacity to assess forest resources;

- support completion of studies of ocean disposal of radioactive waste;

support a ban on the export of hazardous waste to countries that do not have the technological capacity to manage it;

encourage the formulation of new regional agreements to protect the marine environment; and

participate fully in the UN program of "prior informed consent" for international shipments of toxic chemicals.[10]

Although Agenda 21 did not establish priorities or provide funds (the estimated cost of implementation was $125 billion) and is not legally binding, it enumerates what delegates considered the most pressing global problems.[11]

The Earth Summit culminated in two global conventions and two other agreements: a climate convention aimed at encouraging nations to reduce emissions of greenhouse gases; a convention aimed at protecting the world's biodiversity by having the wealthy nations fund efforts to protect endangered species and ecosystems and by regulating biotechnology; an agreement to limit deforestation and logging in tropical forests; and an agreement by the wealthy nations to provide $2 billion in additional assistance to the poorer nations. Despite the expansive agenda, however, the Earth Summit failed to address a number of major issues. Little discussion focused on direct efforts to address global hunger, malnutrition, poor health, poverty, drought, and agricultural problems, for example.[12] Undoubtedly, however, there is some value in preparing such documents as Agenda 21 and the Rio Declaration. These documents focus attention on problems, provide an opportunity to redirect policy by those disposed to do so, and can serve as the basis for further negotiations and eventual agreements.

Follow-up from the 1992 Earth Summit has been sluggish, but some seeds of change are taking root. The Earth Charter will be presented to the UN in 1997 for adoption by 2000. If adoption is the result of a worldwide process of discussion and deliberation, the document may carry sufficient authority to compel governments to

International Treaties and Other Legal Agreements

comply with at least some of its provisions. Many communities have launched their own Agenda 21 platforms of action for ecologically sustainable economic activity. National councils for sustainable development, composed of governments and elements of civil society, have been proposed and some experiments are taking place. Within their professional associations, engineers, architects, and others have formed groups aimed at promoting sustainable development.

This chapter describes the key provisions of the major international environmental agreements, the obligations that parties to those agreements accept, and the implications of their efforts to implement the provisions. The chapter focuses on four major areas of concern: the atmosphere—including acid rain, the ozone layer, and climate change—marine resources, biodiversity, and population growth.

International Treaties and Other Legal Agreements

Some 175 international agreements regarding environmental problems are currently in effect, on issues ranging from fisheries to the conservation of plant life to ocean pollution and development. Among the earliest treaties were two North American agreements, the 1909 Boundary Waters Treaty concluded between Great Britain (on behalf of Canada) and the United States, and the 1906 Convention Concerning the Equitable Distribution of the Waters of the Rio Grande for Irrigation (U.S.–Mexico).[13] Several important regional agreements, particularly those fashioned under the auspices of UNEP's regional seas program, have been signed more recently. Box 1.1 lists the major global environmental accords by broad subject area. This is only a sampling of the treaties that have been negotiated. Many of them involve complex packages of general agreements and specific protocols and appendices. But the agreements demonstrate optimism that the nations of the world can come together to solve common environmental concerns.

> **BOX 1.1**
>
> **SELECTED MAJOR INTERNATIONAL ENVIRONMENTAL AGREEMENTS AND THE SECRETARIATS RESPONSIBLE FOR ADMINISTERING THEM**
>
> ■
>
> **Pollution of the Atmosphere**
>
> 1979 Convention on Long-Range Transboundary Air Pollution (LRTAP)—UN and Economic Commission for Europe (ECE)
> 1985 Helsinki Protocol Concerning the Reduction of Sulphur Emissions or Their Transboundary Fluxes
> 1985 Vienna Convention for the Protection of the Ozone Layer—UNEP
> 1987 Montréal Protocol on Substances that Deplete the Ozone Layer, London Amendments, 1990
> 1988 Sofia Protocol Concerning the Control of Emissions of Nitrogen Oxides or Their Transboundary Fluxes
> 1991 Geneva Protocol Concerning the Control of Emissions of Volatile Organic Compounds or Their Transboundary Fluxes
> 1992 UNCED Framework Convention on Climate Change—UNEP
>
> **Marine Pollution and Resources**
>
> 1946 Convention for the Regulation of Whaling
> 1954 Convention for the Prevention of Pollution of the Sea by Oil
> 1958 Convention on the High Seas
> 1958 Convention on the Continental Shelf
> 1958 Convention of the Territorial Sea and the Contiguous Zone
> 1958 Convention of Fishing and Conservation of the Living Resources of the High Seas
> 1969 Agreement for Cooperation in Dealing with Pollution of the North Sea by Oil
> 1972 Convention on the Prevention of Marine Pollution by Dumping of Wastes and Other Matter (London Dumping Convention)—International Maritime Organization (IMO)

Treaties are written agreements between two or more states (nations); *conventions* are multilateral treaties, but these terms are often used interchangeably. *Protocols* are agreements that amend or supplement a treaty or convention. There are different levels of

International Treaties and Other Legal Agreements

1973   International Convention for the Prevention of Pollution from Ships and the Protocol of 1978 Relating Thereto with Annexes (MARPOL)—IMO
1974   Convention for the Prevention of Marine Pollution from Land-Based Sources (Paris Convention)
1974   Convention on the Protection of the Marine Environment of the Baltic Sea Area (Helsinki Convention)
1976   Convention on Protection of the Rhine Against Chemical Pollution
1976   Convention for the Protection of the Mediterranean Sea Against Pollution
1982   UN Convention on the Law of the Sea (UNCLOS)—UN Office for Ocean Affairs and the Law of the Sea (not in force)
1990   International Convention on Oil Pollution Preparedness, Response and Cooperation—IMO

Hazardous Substances

1986   Convention on Early Notification of a Nuclear Accident—International Atomic Energy Agency (IAEA)
1986   Convention on Assistance in the Case of a Nuclear Accident or Radiological Emergency—IAEA
1989   Convention on the Control of Transboundary Movements of Hazardous Wastes and their Disposal (Basel Convention)—UNEP
1991   Convention on the Ban of the Import into Africa and the Control of Transboundary Movements and Management of Hazardous Wastes within Africa (not in force)

Species and Habitat

1911   Convention for the Preservation and Protection of Fur Seals
1940   Washington Convention on Nature Protection and Wild Life Preservation in the Western Hemisphere—Organization of American States
1946   International Convention for the Regulation of Whaling—International Whaling Commission
1950   Paris International Convention for the Protection of Birds
1951   International Plant Protection Convention

(Continued on page 22)

participation in treaty or convention making: a *signatory* is a nation that has signed an agreement, but simply signing creates no binding obligations; a *party* to an agreement is a state that has ratified it, following whatever domestic process its laws require. When the

(BOX 1.1 *Continued from page 21*)

1954 Convention for the Prevention of Pollution of the Sea by Oil—International Maritime Organization
1959 Antarctic Treaty
1963 Treaty Banning Nuclear Weapons Tests in the Atmosphere, in Outer Space and Under Water
1971 Ramsar Convention on Wetlands of International Importance Especially as Waterfowl Habitat—World Conservation Union
1972 Paris Convention Concerning the Protection of the World Cultural and Natural Heritage (World Heritage Convention)—United Nations Educational, Scientific, and Cultural Organization (UNESCO)
1973 Washington Convention on International Trade in Endangered Species of Wild Fauna and Flora (CITES)—United Nations Environment Program
1979 Bonn Convention on the Conservation of Migratory Species of Wild Animals—UNEP
1980 Convention on the Conservation of Antarctic Marine Living Resources—CCAMLR Secretariat
1983 International Undertaking on Plant Genetic Resources—Food and Agriculture Organization
1991 Protocol to the 1959 Antarctic Treaty on Environmental Protection (not in force)
1992 UNCED Convention on Biological Diversity—UNEP

Other

1992 UNCED Agenda 21
1994 Program of Action, International Conference on Population and Development, Cairo

NOTE: The list above includes only some of the major international accords. A number of important regional agreements, particularly those fashioned under the auspices of UNEP's regional seas program, have been signed, but are not listed here.
SOURCE: Adapted from Lee A. Kimball, *Forging International Agreement* (Washington, D.C.: World Resources Institute, 1992), 75–76; and Lynton Keith Caldwell, *International Environmental Policy: Emergence and Dimensions*, 2d ed. (Durham, N.C.: Duke University Press, 1990), 349–51.

requisite number of states ratify an agreement, as defined in that agreement, it *enters into force* and its provisions become binding.

Environmental treaties and conventions typically include the following kinds of provisions: (1) goals to reduce or eliminate pollution, ensure environmental quality, protect resources, and so forth, and timetables by which those goals are to be achieved; (2) standards the parties are to enforce within their own borders; (3) requirements for permitting or licensing systems administered by national governments to ensure compliance by private and subnational governmental actors; (4) obligations to notify other parties of dangerous releases or other problems; (5) submission of data demonstrating compliance to an international secretariat; (6) creation of a secretariat to monitor progress and facilitate future agreements; (7) dispute-resolution mechanisms in the event of disagreements over compliance by parties; and (8) sanctions, incentives, and penalties to encourage compliance by the parties, and, in particular, assisting the less developed nations meet their obligations.

Agreements to Protect the Atmosphere

Nuclear Testing in the Atmosphere

The six nations possessing nuclear weapons—the United States, Russia, France, Britain, China, and India, have conducted at least 528 atmospheric tests of nuclear devices.[14] The 1963 partial test ban treaty prohibiting atmospheric, ocean, or space testing of nuclear weapons has been successful: There have been no known atmospheric tests since 1980, and more than one hundred nations are parties to the agreement. The treaty can be viewed as an early example of the "precautionary principle": there was considerable uncertainty concerning the environmental and health risks posed by low-level exposure to radioactive fallout, but widespread concern generated pressure for limits on atmospheric testing.[15]

Achieving a comprehensive nuclear test-ban treaty has been more difficult. In 1995, representatives of more than one hundred countries joined in negotiations for a treaty to end all testing of weapons; by August 1996, the parties had agreed on a draft agreement. India then announced it wanted to delay negotiations until a timetable could be reached for eliminating all nuclear weapons; skeptics believed India wanted to delay the negotiations until it could continue its own testing program.[16] The treaty languished until Australia brought the issue before the UN General Assembly. Because of its credibility among the nonaligned nations, Australia was able to rally support for the agreement. In September 1996, the UN General Assembly voted overwhelmingly (158–3) to approve the treaty, preparing the way for its signing and ratification.[17]

Except for India, the existing nuclear powers all announced their support for the treaty.[18] However, India's decision to oppose it means that the treaty's international monitoring and inspection system cannot be put in place. The treaty requires all forty-four countries that have nuclear reactors to sign and ratify the treaty before it becomes international law. Informal enforcement may nevertheless occur as such countries as the United States detect violations using their own intelligence agencies and publicize those violations in an effort to put pressure on the offending nation. The treaty also provides that a review conference be convened within three years to bring the treaty into force if the requisite nations have not signed and ratified the accord.

Acid Rain

Acid deposition, or acid rain, is a by-product of the burning of fossil fuels, which produces sulfur dioxide (SO_2) and nitrogen oxides (NO_x). These gases are transformed in the atmosphere into sulfuric acid and nitric acid. The acids usually remain in the atmosphere for weeks and may travel hundreds of miles before settling back to earth as dry particles or precipitation. Acidic particles and gases formed at ground level are directly absorbed by plants or are oxidized into sulfates and nitrates and absorbed into the soil.

Acid rain has been of greatest concern in Europe and North America. According to one estimate, 75 percent of the commercial forests in Europe have suffered damage from air pollution, primarily from acid rain, but also from ground-level ozone. Acid deposition also damages stone and metal in buildings and monuments.[19] Most of the concern surrounding acid rain in the United

States has focused on its impact on forests and aquatic resources.[20] Reducing the damage caused by acid rain is by itself complicated enough. The threat of global warming and climate change has caused this goal to become intertwined with efforts to control carbon dioxide and methane emissions and to decrease the use of fossil fuels.[21]

Several international agreements and national regulatory programs have been fashioned to reduce the risk of acid rain to human health and ecosystems (see Box 1.2). Thirty-five nations from Europe and North America signed the 1979 Convention on Long-

---

BOX 1.2

ACID RAIN AGREEMENTS

■

Convention on Long-Range Transboundary Air Pollution, Geneva, 1979; entered into force for the United States on March 16, 1983. Provisions seek to protect humans and their environment from air pollution; limit and, as far as possible, gradually reduce and prevent air pollution, including long-range transboundary air pollution. It is open to member States of the United Nations Economic Commission for Europe (UN/ECE), the European Economic Community, and other European States having consultative status with the UN/ECE; forty states are parties, including the European Economic Community. The Secretariat is the UN/ECE Environment and Human Settlements Division (ENHS), Geneva, Switzerland.

Protocols include:

Protocol on Long-Term Financing of the Cooperative Programme for Monitoring and Evaluation of the Long-Range Transmission of Air Pollutants in Europe, Geneva, 1984, entered into force on January 28, 1988

1988 Protocol to Freeze Nitrogen Oxide Emissions at 1987 Levels by 1994; will enter into force ninety days after sixteen nations ratify or accept; United States accepted on July 13, 1989

1991 Canada–United States Air Quality Agreement: both nations agree to reduce sulfur dioxide emissions by specified amounts.

NOTE: Figures for the number of parties of this and other agreements discussed in this chapter are as of October 31, 1995.
SOURCE: Fridtjof Nansen Institute, *Green Globe Yearbook of International Co-operation on Environment and Development* (New York: Oxford University Press, 1996), 96–97.

Range Transboundary Air Pollution, which established a framework for subsequent negotiations.[22] In 1985 most of these nations, although not the United States, agreed to cut sulfur dioxide emissions, the primary precursor of acid rain, by 30 percent from 1980 levels by 1993. In 1988, a second agreement placed a cap on nitrogen oxide emissions, the other major contributor to acid deposition, at 1987 levels.[23]

Sulfur dioxide emissions fell dramatically in western Europe from 1980 to 1990 (Austria by 75 percent, Belgium by 49 percent) under a 1979 treaty on transboundary air pollution and subsequent tightening measures. However, these reductions have not been great enough to protect some ecosystems adequately; damage to plant and aquatic life continues.[24]

ACID RAIN AGREEMENTS BETWEEN THE UNITED STATES AND CANADA

The evolution of acid rain policy in North America illustrates well some of the challenges of formulating and implementing bilateral environmental agreements. Acid rain became one of the most highly publicized environmental problems of the 1980s and a major source of contention between the United States and Canada. The two countries began informal discussion of alternatives for reducing acid rain in 1978, and a Memorandum of Intent was signed in 1980 to begin formal negotiations. After three years of talks, Canada proposed that both countries reduce sulfur dioxide emissions by 50 percent, but the Reagan administration rejected the proposal, claiming that the United States had already done much more than Canada to reduce sulfur emissions and that more research was necessary before making major investments in pollution control technologies.[25] Canadians argued that they had previously taken significant steps to reduce their contribution to the problems of acid rain and challenged the United States to do the same. In 1985 they created a plan to reduce sulfur dioxide emissions in 1994 by one-half of their 1980 levels and to reduce emissions by 67 percent in Ontario. Special envoys from the two nations were appointed in 1985 to study acid rain; they focused in particular on the devel-

Agreements to Protect the Atmosphere

opment of cleaner coal-burning technologies. Negotiations eventually resulted in an agreement in 1986 that called for the United States to spend $5 billion over five years, funded equally by industries and the federal government, to retrofit plants with new control technologies.[26]

Sulfur dioxide and nitrogen dioxide emissions have been regulated in the United States for more than two decades under the Clean Air Act. For these two pollutants, the EPA has issued National Ambient Air Quality Standards that are designed to protect human health in specific air quality regions. Since the 1970 and 1977 Clean Air Acts focused on local air pollution, the law required a major amendment to address acid rain.

The domestic politics of acid rain made the enactment of legislation much more complicated in the United States than in Canada. The Reagan administration and congressional opponents argued that there was insufficient information to link emissions from power plants and factories in the Midwest with acid rain damage in Canada and New England. In any event, they argued, the problem would be remedied as old power plants were replaced by new, less polluting ones.[27] Several proposals were made to regulate acid rain throughout the 1980s. The impasse was broken when the Bush administration released its proposed amendments to the Clean Air Act in the summer of 1989, and the debate over whether new acid rain controls were necessary was essentially concluded.[28] One of the most important innovations of the 1990 Clean Air Act was the creation of pollution allowances that sources could buy and sell. It helped break the deadlock blocking passage of the bill and may serve as the forerunner for a new generation of environmental laws that rely on market-based approaches.[29]

THE 1991 CANADA–UNITED STATES AIR QUALITY AGREEMENT

President Bush met with Prime Minister Mulroney in February of 1989 to begin discussing an acid rain agreement. Negotiations continued throughout 1989 and 1990 as the Clean Air Act wound its way through Congress. The talks culminated in an Air Quality Agreement that was signed on March 13, 1991. In the agreement, Canada agreed to a national cap of 3.2 million tonnes of sulfur dioxide emissions by the year 2000. The United States essentially agreed to make the reductions mandated in the 1990 Amendments to the Clean Air Act (see Table 1.1).

The differences in the approaches taken by the two countries is noteworthy. Policy makers in both countries are increasingly interested in marketlike incentives in regulation as a way to reduce compliance costs.[30] Acid rain legislation in the United States is the first major legislative test of such an approach. The key to the acid rain emissions trading system is the idea of a cap; the actual distribution of emissions is viewed as less important. It is not clear how such a scheme could work in achieving the ambient air quality standards for criteria pollutants, or for regulating air toxics, since the distribution of these pollutants is critical. If air quality in specific areas does not meet a certain standard, then the threat to the health of local residents is a problem, even if nationwide levels are reduced. Even within a facility, trading may not work if it means that more dangerous emissions can be substituted for less dangerous emissions (see Chapter 3).

In contrast, the small number of sources precluded the creation of an emissions trading system in Canada. The Canadian structure

TABLE 1.1

SULFUR DIOXIDE TRENDS AND TARGETS

Year	Canada Million Tonnes*	United States Million Tons
1970	6.9	31.2
1980	4.6	25.7
1990	3.7	23.3
2000	3.2	16.3
2010	3.2	15.7

*1 (metric) tonne = 1.1 short, or U.S., tons (2,200 pounds).
SOURCE: "Canada/United States Air Quality Agreement, Progress Report" (March 1992).

of government is also more decentralized and provinces have much more discretion. Canadian provisions have traditionally had inadequate monitoring and compliance systems that would also make a market-based system problematic. The acid rain agreement demonstrates the need to combine traditional command-and-control regulation with economic-based initiatives, and will provide some lessons for possible ways of implementing global climate change accords (see below).

The Stratospheric Ozone Layer

At low altitudes, ozone is a noxious pollutant. In the stratosphere, however, ozone provides a protective layer that absorbs most of the ultraviolet rays from the sun. Ultraviolet radiation can, when it reaches the earth, damage biological molecules, including DNA, and increase the incidence of skin cancer, cataracts, and immune deficiencies. It can also harm crops and damage fragile aquatic ecosystems. In short, without the protection which the ozone layer provides, life as we know it on earth would probably not exist at all.[31]

In the 1960s scientists first found evidence that the ozone layer was being destroyed. Scientists believed this destruction was generally attributable to solar ultraviolet radiation, which, during periods of high intensity, was thought to bombard ozone molecules and break them down. As this radiation lessened, the ozone would regenerate itself. In 1971, researchers became concerned that the proposed fleet of supersonic transport (SST) aircraft would damage the ozone layer because emissions of water vapor and nitrogen oxides from the SSTs had been shown to attack ozone. In later years, rising levels of nitrous oxides (N_2O) from increased combustion and increased use of nitrogen-rich fertilizers led to similar concerns that the ozone layer was being destroyed. In 1974, two scientists discovered a new threat to the ozone layer, one which overshadowed all previous concerns. Mario J. Molina and F. Sherwood Rowland proposed that the increasing use of compounds known as chlorofluorocarbons (CFCs), used in refrigerants and aerosol sprays, posed a serious threat to the ozone layer.[32]

INTERNATIONAL ENVIRONMENTAL AGREEMENTS

Research in the mid-1970s led to the first international agreement to protect the ozone layer (see Box 1.3). In 1978, the United States, Canada, and several Scandinavian countries agreed to restrict or ban nonessential aerosol uses of CFCs. By 1983, regulatory action and voluntary industry restrictions in several CFC-producing countries resulted in a 21 percent reduction of CFC from the peak production level of 1974.[33]

In May 1981, a working group was established under the United Nations to outline a framework for a general convention on the protection of the ozone layer. Despite the failure to agree on a specific

---

BOX 1.3

AGREEMENTS TO PROTECT THE OZONE LAYER

■

Agreement Regarding Monitoring of the Stratosphere (three-party agreement among the United States, the United Kingdom, and France), Paris, 1976; entered into force for the United States on May 5, 1976.

Vienna Convention for the Protection of the Ozone Layer, 1985; United States ratified on April 21, 1988, entered into force for the United States on November 22, 1988. The agreement calls for measures to control human activities found to have adverse effects on the ozone layer; cooperation in scientific research and systematic observations of the ozone layer; and exchange of information. It is open to all states. There are one hundred and fifty parties to the Montréal Protocol, including the European Economic Community. The secretariat is UNEP, Ozone Secretariat, Nairobi, Kenya.

Protocols include:

> Montréal Protocol to the Vienna Convention for the Protection of the Ozone Layer on Substances that Deplete the Ozone Layer, 1987; ratified by the United States in March 1988; entered into force on January 1, 1989.
>
> London agreement to ban ozone-depleting substances, 1990; entered into force on August 10, 1992; one hundred and three parties; specified twelve new chemicals to be controlled and thirty-four new chemicals to be reported; and established a multilateral fund to assist parties with compliance costs.
>
> Copenhagen agreement to establish multilateral fund, 1992; forty-nine parties; called for earlier phase-out of some substances and confirmed arrangements for the multilateral fund.

Agreements to Protect the Atmosphere

provision for CFCs, the draft convention of the working group was adopted in 1985 as the Vienna Convention for the Protection of the Ozone Layer and was signed by twenty nations. The major provisions of the agreement outlined the general responsibility of states for preventing environmental harm caused by human interference with the ozone layer; specified the duties of states for intergovernmental cooperation, including monitoring, information exchange, and the harmonization of national measures; and created new international institutions (including a conference of the parties to the convention and a secretariat) to implement the agreement and to

The current regulatory requirements for these chemicals are as follows:

Chemical	Lifetime in atmosphere	Production banned by
Chlorofluorocarbons (CFCs) (used in refrigerators and air conditioners and as solvents and sterilants)	100	1/1/1996
Halons (used in fire extinguishers)	100	1/1/1994
Carbon tetracholoride (used in air conditioners, plastic insulation, and packaging foam)	50	1/1/1996
Methyl chloroform	6	1/1/1996
Methyl bromide	–	Production freeze at 1991 levels

NOTE: Essential uses of CFCs, halons, carbon tetrachloride, and methyl chloroform are exempted from the ban. A 99.5 percent phase-out of HCFCs is projected to occur by 2020. Methyl bromide production is to be reduced to 1991 levels by 1995.

SOURCE: Fridtjof Nansen Institute, *Green Globe Yearbook of International Co-operation on Environment and Development* (New York: Oxford University Press, 1996), 104–08; Council on Environmental Quality, *Environmental Quality* (Washington, D.C.: CEQ, 1993), 139.

develop more specific rules in the form of protocols, technical annexes, and financial and procedural provisions.[34]

The Vienna Conference was only the beginning of international efforts to protect the ozone layer. New concerns arose over the accelerated rate at which the ozone layer was being depleted. In late 1985, a British research team published a report stating that springtime amounts of ozone over Antarctica had decreased by more than 40 percent between 1977 and 1984. Other research groups confirmed the British report and showed that the region of depletion was wider than the continent of Antarctica and spanned much of the lower stratosphere. In effect, what had been found was a "hole" in the earth's protective layer of ozone. The discovery prompted a search for the cause of the ozone loss. Most researchers returned to an explanation proposed over a decade before: that chemicals, specifically CFCs, were depleting the ozone layer. Laboratory studies had previously showed that chlorine destroys ozone. The millions of tons of CFCs that had been released into the atmosphere for decades were sufficient to produce the kind of damage that had been identified.

THE MONTRÉAL PROTOCOL

In 1986, delegates from fifty-four nations met in Geneva, Switzerland, to discuss limitations in the production and use of CFCs. The Montréal Protocol eventually resulted from the Geneva meeting. The Montréal agreement called for a 50 percent cut in CFC production by the year 2000. However, events changed the timetable. A NASA expedition brought back definitive proof of a hole in the ozone layer, thus strengthening the arguments against CFCs. This new evidence, coming barely a month after the signing of the protocol, demonstrated the inadequacy of its provisions and generated renewed calls for a total ban of CFCs. Findings released in 1988 by the Global Ozone Trends Panel, an international group of more than one hundred scientists, concluded that ozone depletion was occurring at two to three times the rate predicted by computer models. This meant, according to the scientists, that the world had already suffered more ozone depletion than was predicted for the

Agreements to Protect the Atmosphere

year 2050 under the terms of the Montréal Protocol.[35] In 1990, eighty-six nations agreed to go beyond the terms of the Montréal Protocol and eliminate halons by 1994 and CFCs by the year 2000.[36] Significantly, the Western industrialized countries and Japan agreed to help the Third World nations achieve the reduction. In 1992, the European Community and the United States, prompted by reports that the ozone layer was thinning more rapidly than had been expected, committed to phase out CFC production by the end of 1995; the 105 developing countries agreed to ban CFC production by 2010.[37]

The experience of the United States in developing legislation to protect the ozone layer is instructive. The 1990 Clean Air Act's provisions, aimed at preventing further depletion of the stratospheric ozone layer, were among the most environmentally aggressive elements of the bill. Unlike other provisions, the ozone provisions required that certain substances be eliminated rather than merely regulated. Opposition centered on the argument that American industry should not be forced to make greater emission reductions than their competitors. Some argued that the ability of the United States to foster future international agreements would be undermined by unilateral action. Industry groups such as the National Association of Manufacturers and the Alliance for Responsible CFC Policy argued against tougher domestic reduction schedules without corresponding international action. "Any unilateral action taken by the United States," the Alliance contended, "would have an all but insignificant effect upon the global environment and severely hinder U.S. industry while placing the American economy at an unfair disadvantage in the global market, to say nothing of the loss in American jobs."[38]

However, support for protecting the ozone layer was widespread, and accounts had already surfaced of industry's ability to find alternatives to CFCs. Both houses of Congress took up the cause of more aggressive protection of the ozone layer, ensuring the passage of the provisions. Subsequent research has identified an even further decline in the ozone layer than was apparent in 1990. If an agreement to reduce the manufacturing and use of CFCs had been reached a decade ago in response to initial scientific warnings, much less damage to the ozone layer would have occurred.

The Montréal Protocol is often cited as an example of a successful global response to an environmental threat and a model of global environmental diplomacy. Former UNEP director Mostafa Tolba hailed it as "the beginning of a new era of environmental statesmanship."[39] Richard Elliot Benedick's study of the agreement argued that it "broke new ground in its treatment of long-term risks and in its reconciliation of difficult scientific, economic, and political issues." According to Benedick, a key figure in the negotiations,

> the Montréal Protocol was the result of research at the frontiers of science combined with a unique collaboration between scientists and policy makers. Unlike any previous diplomatic endeavor, it was based on continually evolving theories, on state-of-the-art computer models simulating the results of intricate chemical and physical reactions for decades into the future, and on satellite-, land-, and rocket-based monitoring of remote gases measured in parts per trillion. An international agreement of this nature could not, in fact, have occurred at any earlier point in history.[40]

One of the most important elements of the ozone protection agreements was the creation of a multilateral fund to help developing countries finance their implementation of the international commitments. The fund was initially endowed with $3 billion from the wealthy countries when it was first created in 1994. This fund is discussed in more detail in chapter 4. Since then, a number of countries have implemented laws and regulations to ensure reduction in the manufacturing or use of CFCs.

In some ways, the threats posed by CFCs are manageable because the number of producers is limited and CFC substitutes have been developed. However, some of those substitutes, including hydrochlorofluorocarbons (HFCFs), hydrofluorocarbons (HCFs), and methyl bromide, are also feared to be ozone depleters. The ban

Agreements to Protect the Atmosphere

on the production of Freon, the most commonly used CFC, has driven the price of existing stocks from $1 a pound in the 1980s to $35 a pound in 1996. That has led to smuggling of Freon into the United States and a rash of CFC thefts. And CFC-containing products that wealthy nations have banned are being sold in the less-developed world. Because of the long time period CFCs remain in the atmosphere, we may be living with the consequences of a depleted ozone layer for decades even though CFC production is declining.

Global consumption of CFCs decreased from a peak of 1.2 billion kilograms in 1987 to some 682 million kilograms in 1991. Scientists reported in 1993 that the buildup of ozone-depleting chemicals began to slow down and was expected to peak around the beginning of the next century. In 1996, scientists at the National Oceanic and Atmospheric Administration began detecting a reduction in ground measurements of chlorine, the chemical that is most responsible for ozone layer depletion. It takes two or three years for the chemicals at ground level to float to the stratosphere; as a result, the peak of ozone layer depletion is expected to occur somewhere between 1997 and 1999. There may be some signs of recovery of the ozone shield by 2005 or 2010. But that optimistic scenario requires continued compliance with the Montréal Protocol and amendments.[41] The participation of the developing countries is critical, since increased use of CFCs in China and India could largely negate reductions in emissions throughout the rest of the world.[42]

Despite the consensus concerning the threat of ozone depletion, the relatively small number of CFC producers, and other factors that should have led to rapid and effective international action, the Protocol was still slow in coming (the agreement was signed thirteen years after publication of the first article that suggested the ozone layer was threatened) and slow in being implemented. Two hundred thousand additional deaths from skin cancer in the United States alone could occur during the next fifty years as a result of damage to the ozone layer.

The challenges in regulating ozone-depleting substances serve as a sobering reminder that dealing with other global threats, such as climate change, will be all the more difficult.[43] We may have acted too late to prevent widespread harm from a weakened ozone layer.

Global Climate Change

Global warming is perhaps the broadest and potentially most catastrophic environmental predicament confronting humans; it is also the most controversial. The gases that contribute to the greenhouse effect are not major constituents of the atmosphere: They comprise less than 1/10 of 1 percent of the atmosphere, except water vapor, which makes up about 1 percent. Human behavior has little impact on levels of water vapor. While the threat of global warming is often equated with rising levels of carbon dioxide, the problem is more complicated. Carbon dioxide and methane are the two most important greenhouse gases; carbon dioxide represents about half of the human-caused contribution to the greenhouse effect. Other gases, particularly methane, nitrous oxide, low-level ozone, and chlorofluorocarbons also trap radiant heat from the earth.[44]

Methane, a natural by-product of bacterial activity in plants and animals (particularly rice and livestock) and of fossil-fuel burning, represents a little less than one-fifth of the marginal or human-caused increase in warming capacity. However, CFCs are the fastest-growing greenhouse gas. They currently represent 15 to 20 percent of the total greenhouse gases; emissions are growing at about 7 percent a year. (Carbon dioxide emissions are increasing annually at about 0.5 percent.) Although the concentration of CFCs in the atmosphere is quite small compared with carbon dioxide levels, each unit of CFCs is much more effective than carbon dioxide in trapping radiant heat from the earth, so reduction of emissions of these gases must also be part of a global agreement to reduce global warming. Nitrous oxide, also produced by bacterial decay in animal wastes and nitrogen fertilizer, makes up about 6 percent of the greenhouse gases. Low-level ozone, produced by some natural releases but primarily through the interaction of sunlight with volatile organic compounds (from chemical manufacturing, petroleum refining, evaporation of solvents, and motor vehicle emissions) and nitrogen oxides (primarily from fossil-fuel combustion), might contribute as much as 10 percent.[45]

Since 1850, the level of carbon dioxide in the atmosphere has increased by about 20 percent, and earth's average temperature has increased by about 1 degree. If this trend continues, the temperature could increase by an additional 2.5 to 5 degrees by the year 2050, about the same level of change in temperature that occurred during the end of the last ice age some fifteen thousand years ago, resulting in massive alteration of the ecosphere.[46]

The threat of global warming grows each year. Fossil-fuel burning releases about 6 billion tons of carbon into the atmosphere each year; 3 billion tons accumulate, and the balance is absorbed by oceans and forests. By 1995, an increase of 170 billion tons of carbon dioxide had accumulated in the atmosphere since the beginning of the industrial revolution.[47]

Many proposals for a global warming agreement have argued for a reduction in greenhouse gas emissions of from 20 to 50 percent of current levels. The 1988 Toronto Conference on the Changing Atmosphere, for instance, produced a call for a 20 percent reduction in world CO_2 emissions by the year 2005, and other groups have recommended similar goals.[48] The 1990 Interparliamentary Conference on the Global Environment, composed of parliamentarians from thirty-five countries, called on nations to commit to a 50 percent reduction in greenhouse gases from 1990 levels by the year 2010.[49] An international workshop convened by the Climate Institute in February 1990 brought together eighty participants from a dozen countries to draft a global warming framework convention. The participants recommended that the industrialized nations to reduce their fossil fuel emissions by 20 percent from 1988 levels by the year 2000 and the developing countries reduce emissions per unit of gross national product by 2.5 percent each year through the year 2000.[50]

THE FRAMEWORK CONVENTION ON CLIMATE CHANGE

The agreement ultimately signed by most of the world in 1992 was much less ambitious (see Box 1.4). The United Nations Framework Convention on Climate Change, one of the agreements resulting from the Earth Summit, calls on the parties to stabilize the concentrations of greenhouse gases in the atmosphere

> at a level that would prevent dangerous anthropogenic interference with the climate system . . . within a time frame sufficient to allow ecosytems to adapt naturally to climate change, to ensure that food production is not threatened and to enable economic development to proceed in a sustainable manner.[51]

BOX 1.4

FRAMEWORK CONVENTION ON CLIMATE CHANGE (FCCC)

The convention was adopted in New York in May 1992, and presented at the 1992 UNCED meeting in Brazil. It commits parties to stabilize greenhouse gas concentrations in the atmosphere to prevent dangerous anthropogenic interference with global climate; to ensure that ecosystems have the time to adapt naturally to any climate change; and to enable economic development to proceed in a sustainable manner. There are 145 parties to the agreement, including the European Community.

The developed countries agreed to:

- take the lead in combating climate change;
- reduce emissions to 1990 levels by the year 2000; and
- provide new and additional financial resources to meet the agreed full costs incurred by the developing countries.

All parties agreed to:

- develop and publish inventories of emissions and sinks of greenhouse gases;
- transfer technologies that control, reduce, or prevent greenhouse gas emissions; and
- take climate change into account in their social, economic, and environmental policies.

Agreements to Protect the Atmosphere

All parties are required to publish regularly "national inventories of anthropogenic emissions by sources and removals by sinks of all greenhouse gases"; to formulate and implement national and, where appropriate, regional plans to "mitigate climate change"; and to consider the threat of climate change in "relevant social, economic and environmental policies and actions."[52] (Carbon sinks, such as forests, absorb CO_2 from the atmosphere.) The developed countries agreed to limit greenhouse gas emissions (of gases subject to the Montréal Protocol); to protect carbon sinks and reservoirs "by the end of the present decade . . . to their 1990 levels"; and to "provide new and additional financial resources to meet the agreed full costs incurred by the developing country Parties in complying with their obligations."[53] The convention provides no specific

---

The accord is open to members of the United Nations, any of its specialized agencies, parties to the Statute of the International Court of Justice, and regional economic integration organizations. The secretariat is the Climate Change Secretariat, Geneva, Switzerland.

The following institutions are involved in climate change:

FCCC Conference of the Parties: meets yearly; adopts amendments and protocols; oversees implementation

Secretariat: provides administrative functions

Subsidiary Body for Scientific and Technical Advice: assists in evaluating scientific and policy questions

Subsidiary Body for Implementation: reviews national reports and assesses implementation efforts

Inter-governmental Panel on Climate Change: established in 1988 by the World Health Organization and the United Nations Environment Program to assess scientific and policy issues

Global Environment Facility: established in 1991 by the World Bank, UNDP, and UNEP; temporary funding agent for the FCCC

SOURCE: Fridtjof Nansen Institute, *Green Globe Yearbook of International Co-operation on Environment and Development* (New York: Oxford University Press, 1996), 101–103.

targets or timetables and makes the effort voluntary; the only enforcement mechanism provided is an agreement by the participating states to report periodically on their progress in achieving the target. For the FCCC to be viable, parties must agree to comply with at least the year-2000 goal and expand monitoring of greenhouse gas emissions; the World Bank and other multilateral institutions will need to increase funding for energy efficiency and related projects.

The prefatory language of the Framework Convention on Climate Change, signed at the 1992 United Nations Conference on Environment and Development (UNCED) in Brazil, emphasized the differences between the less- and more-developed nations. The parties recognized that the "largest share of historical and current global emissions of greenhouse gases has originated in developed countries, that per capita emissions in developing countries are still relatively low and that the share of global emissions originating in developing countries will grow to meet their social and development needs." Efforts to respond to the threat of climate change must be "coordinated with social and economic development in an integrated manner with a view to avoiding adverse impacts on the latter, taking into full account the legitimate priority needs of developing countries for the achievement of sustained economic growth and the eradication of poverty." The developing countries "need access to resources required to achieve sustainable social and economic development."

The implementation of the provisions of the Convention is primarily the responsibility of the participating states, but the agreement created a "Conference of the Parties" to "keep under regular review the implementation of the Convention" and "promote the effective implementation of the Convention" and a subsidiary body to "assist the Conference of the Parties in the assessment and review of the effective implementation of the Convention." Each developed-country party is to submit a "detailed description of the policies and measures that it has adopted to implement its commitment" under the Convention, and a "specific estimate of the effects that the[se] policies and measures . . . will have on anthropogenic emissions by its sources and removals by its sinks of greenhouse gases."

Agreements to Protect the Atmosphere

Developing-country parties may "propose projects for financing, including specific technologies, materials, equipment, techniques or practices that would be needed to implement such projects."[54] The convention also recognized the Global Environment Facility as the financial body responsible for helping fund compliance efforts in the developing world. As of October 1995, there were ninety-six parties to the FCCC; an additional seventy-four countries had signed the agreement but not ratified it.[55]

Implementation of the accord has been controversial. Some parties have delayed their expected compliance date from 2000 until 2005. Others have changed the base date from which reductions are to be made in order to reduce required cuts, promised only to freeze emissions beginning in the year 2000, or proposed that emissions be stabilized on a per capita basis, thus permitting increased emissions as population grows. Some have simply disregarded the treaty. Carbon dioxide emissions in the United States in 1994 exceeded the levels to be achieved by 2000, and in that year Congress approved only half the funds requested to comply with the convention.[56]

The climate change convention was approved by the U.S. Senate in 1992 and was subsequently ratified by the requisite fifty nations. In October 1993, the Clinton administration released its Global Change Action Plan as required by the convention.[57] The plan promises to reduce levels of greenhouse gas emissions in the year 2000 to 1990 levels, about 1.5 billion tons, a reduction of some 110 million tons in 1993 emissions. Progress would be measured every two years.

The plan includes nearly fifty new and expanded initiatives. Some eighteen actions will promote energy efficiency among commercial, residential, and industrial users, including demonstration projects for emerging technologies, the upgrade of energy efficiency standards, and funding for investments in energy efficiency in government buildings. Four programs will reduce energy consumption in transportation by reforming tax expenditures for employer-provided parking, promoting telecommuting, reducing the number of motor vehicle miles traveled, and developing fuel economy labels for tires. Nine programs are aimed at increasing the supply of energy, promoting the use of cleaner fuels, developing new technologies, and increasing the efficiency of energy production. Eight initiatives will reduce methane production or increase its recovery, primarily through research and development programs and regulation of landfills, coal mining, and livestock production. Four programs will reduce use of CFCs, HCFCs, nitrogen oxide, and other emissions from industries and fertilizer use. Four efforts will accelerate tree production and reduce loss of forests.[58] One of the most perplexing challenges confronting the Clinton plan has been to reduce energy consumption in the face of declining prices. The American Petroleum Institute released a study in 1995 which concluded that the price of gasoline, after adjusting for inflation, was the lowest it has ever been since gas prices were first recorded seventy-five years ago. Since 1967, the cost of gas has risen more slowly than any other consumer good.[59]

Opponents of efforts to take preventive action against the risk of global warming are led by the fossil fuel industry, which has a tremendous incentive to protect current patterns of energy production and consumption. Some scientists have emphasized the uncertainty surrounding many issues and have discouraged any premature action.[60] However, a great number of studies, including six National Academy of Science reports and one completed by the International Panel on Climate Control, involving hundreds of authors and reviewers, show a substantial area of agreement.[61] Climatic change has always been part of the earth's natural evolution, but the rate of change that appears to be occurring is extraordinary. If these rates of change are realized, they could result in massive disruptions and dislocations. Making changes of such a magnitude is unprecedented in human history. Technologies for improved energy efficiency and alternative fuels exist, are already being used, and have great potential for reducing carbon dioxide emissions as well as other air pollutants.

Many of the less-developed countries are doing even less to comply with the global climate convention. These countries only contribute about one-third of total CO_2 emissions, although they make up 80 percent of the world's population. Emissions are increasing there by more than 5 percent a year, in contrast with a 1 percent increase in the more-developed world. Emissions from China represent one-half of all emissions in Asia, and plans call for dramatic increases in emissions as its coal resources are increasingly used. Tension over who is responsible for global warming and where the greatest future threats lie poses a serious challenge to collective action. Industrial nations are responsible for the vast majority of emissions, and the control the technology that can solve the problem, but cooperation from the developing countries is essential.[62]

Some of the industrial nations favor joint implementation—agreements between the less-developed countries (LDCs) and more-developed countries (MDCs) to develop projects aimed at reducing greenhouse gas emissions—because it promises to encourage the most cost-effective solutions. Joint-implementation agreements may pave the way for a global emissions trading program (see Chapter 3) for carbon dioxide and/or other greenhouse gases. However, joint implementation may provide an excuse for industries to fund relatively cheap projects abroad and reduce pressure to take action at home. More direct assistance can be provided to the less-developed world through transfer of technology, technical assistance in preparing national climate plans, and funding for conservation and alternative technologies (see Chapter 6).

THE FIRST CONFERENCE OF THE PARTIES TO THE RIO CONVENTION

The First Conference of the Parties to the Rio Convention met in Berlin in March and April 1995. Representatives from one hundred and twenty countries agreed to begin negotiations, to be completed by 1997, for reducing greenhouse gas emissions after the year 2000. The Berlin conference recognized the fact that the commitments made by the developed world to stabilize emissions at 1990 levels would be insufficient to achieve the convention's goals. For the first time, the developing countries took the lead in demanding more aggressive global action for climate change. The Alliance of Small Island States, thirty-six nations that are threatened by a rise in sea level, were joined by India, Brazil, Egypt, China, and other developing countries that have come to believe that their development efforts might be undermined should significant changes in climate occur. Some seventy LDCs (excluding the major oil-producing nations) joined in a "green paper" that called for major reductions in industrial nations' emissions of greenhouse gases. Many MDCs, led by the United States, Australia, and Canada, fought against the proposal.

One of the most interesting developments at Berlin was the involvement of the insurance industry. Insurance companies, already reeling from billions of dollars of claims from weather-related damage and fearful of the magnitude of future claims that might result from climate change, began to lobby for stronger international efforts. Insurance industry officials emphasized the economic and environmental benefits of energy-efficient technologies; companies producing those technologies, led by the Business Council for a Sustainable Energy Future, also helped counter the fossil-fuel industry. The final statement called for substantial reduction of greenhouse gas emissions, revision of national climate plans, and creation of a pilot program for joint implementation efforts through 1999. New limits were not agreed to in Berlin, and there was little agreement over how compliance with any limited could be enforced.[63]

However, in July, 1996, at the Geneva Climate Summit, the United States announced a shift in policy and committed to legally binding targets and timetables for reducing greenhouse gas emissions from the more developed world if other countries made a similar pledge. The release of the intergovernmental Panel on Climate Change's Second Assessment Report, in December 1995, prompted renewed commitments to reducing the threat of global climate change. Many scientists had believed that no definitive links would be found between human activity and climate change until the twenty-first century, but the 1995 report, involving some 2,500 scientists worldwide, concluded that the "balance of evidence suggests a discernible human influence on global climate."[64] The United States based its shift in support for legally binding emissions-reduction targets on the 1995 IPCC report. The delegates from the United States and more than one hundred other countries at the 1996 Geneva Conference endorsed the IPCC's conclusions and agreed to support legally binding commitments to reduce significantly greenhouse gas emissions. The United States rejected as too ambitious the proposal from the small island states to reduce greenhouse gas emissions by 20 percent by the year 2005, but also argued that voluntary commitments to reduce emissions were not working. European officials favored nonbinding targets; the United States offered to support giving states the flexibility to achieve the reductions any way they chose. The negotiations will continue until the Third Climate Summit in December 1997, in Japan.[65]

ASSESSMENT OF THE AGREEMENTS

Despite progress, much still needs to be done to put in place an effective regime for climate change. One immediate need, for example, is to collect more data concerning greenhouse gas emissions, but there are considerable obstacles: The carbon content of coal varies from one source to another by at least a factor of 2; we have little information concerning methane emissions from petroleum production; and animal and cropland production of nitrous oxide is extremely difficult to estimate. The lack of a global monitoring effort will offer opportunities, when binding limits are eventually imposed, to cheat on self-reporting data. Some kind of external mechanism to verify emissions reports seems unavoidable, along the lines of arms control treaties.[66]

Another problem is the appropriateness of the distinction made between the less- and the more-developed nations. The current approach is to divide countries into two groups, developed and developing, and impose different requirements. Even so, between the richest and poorest industrialized nations have great differences, as do the poorest of the poorest and the emerging LDCs. However, it is not clear that this division makes sense: Might it not simply exacerbate conflict? Can an equitable sharing of responsibility be devised?[67] Tremendous differences become apparent in the distribution of the sources of greenhouse emissions when measured in total and per capita terms. Only three countries—the United States, the former Soviet Union, and China—account for about one-half of global CO_2 emissions. Despite progress made in the last thirty years in the United States to reduce the amount of CO_2 emissions per unit of economic activity, Americans still produce more CO_2 per person than the people of any other nation.[68] The level of CO_2 emissions are primarily a function of economic development. For the present, in many developing countries, the primary reason for the growth of carbon dioxide levels in the atmosphere is land-use practices, particularly deforestation. Since population growth rates are so much greater in the developing world than the industrialized nations, carbon dioxide emissions from that part of the world are expected to expand greatly. According to a 1990 report by the United Nations Population Fund, by the year 2025, Third World countries may be producing four times the amount of carbon dioxide they currently emit.[69] Pressure to improve the quality of life in these countries collides with concerns about the consequences of increased economic activity on the global environment. Unless that growth is radically different from current trends, the prospects for preventing global warming are bleak, even if the wealthy countries reduce their emissions.

The top twenty-five producers of greenhouse gases in 1991 are listed in Table 1.2. These figures are based on emissions of carbon dioxide (from fossil fuel emission and deforestation), methane (from landfills, coal extraction, oil and gas production, wet rice agriculture, and livestock), and chlorofluorocarbons (CFC-11 and -12). Emission amounts are then multiplied by the global warming potential to arrive at an overall measure of each country's contribution to potential climate change.[70]

The ranking of countries according to per capita contribution paints a different picture. The world median per capita emissions of greenhouse gases was 2.59 metric tons in 1991, coincidentally the level of emission for each resident of China. Table 1.2 uses the ratio of a country's per capita emissions to this median figure to provide a ranking of per capita emissions. China, Brazil, and India, for example, although large overall contributors, are relatively small per capita contributors. U.S. per capita emissions are nine times those of China and eighteen times those of India.

TABLE 1.2

MAJOR PROCEDURES OF GREENHOUSE GASES AND PER CAPITA EMISSIONS

Rank	Country	% of Global Emissions	Rank	Country	Per Capita Emissions in Tons*
1	United States	19.14	1	Qatar	18.63
2	Former USSR	13.63	2	Gabon	17.03
3	China	9.92	3	United Arab Emirates	16.16
4	Japan	5.05	4	Brunei	11.51
5	Brazil	4.33	5	Luxembourg	11.41
6	Germany	3.75	6	Iraq	10.84
7	India	3.68	7	United States	8.95
8	United Kingdom	2.37	8	Bahrain	8.43
9	Indonesia	1.89	9	Australia	7.70
10	Italy	1.72	10	Bolivia	7.68
11	Iraq	1.71	11	Canada	7.10
12	France	1.63	12	Bulgaria	6.74
13	Canada	1.62	13	Suriname	6.63
14	Mexico	1.43	14	Trinidad/Tobago	6.53
15	Poland	1.16	15	Singapore	6.33
16	Australia	1.13	16	Venezuela	6.01
17	South Africa	1.12	17	Saudi Arabia	5.95
18	Spain	1.01	18	Former USSR	5.68
19	Venezuela	1.01	19	Norway	5.68
20	Republic of Korea	0.98	20	Denmark	5.61
21	Zaire	0.93	21	Germany	5.54
22	Thailand	0.88	22	Czechoslovakia	5.30
23	Korea, DPR	0.88	23	United Kingdom	4.87
24	Iran	0.82	24	Japan	4.80
25	Saudi Arabia	0.78	25	Belgium	4.76

*The top per capita countries are large producers of oil or, in the case of Gabon, have undergone massive deforestation.

SOURCE: World Resources Institute, *World Resources 1994–95* (Washington, D.C.: WRI, 1994), 201–202.

Global climate change may also result in an increase of tropical diseases such as malaria, dengue fever, and yellow fever; as temperatures rise, they could spread to more temperate latitudes or to areas that are now too dry to host them.[71] Warming may also threaten the boreal forests that surround the Arctic Circle. These forests, about one-third of the planet's total forested areas, are particularly sensitive to climate change. Warmer weather could increase pest outbreaks and fires, and cause the decline of some species. Boreal forests may also be the largest terrestrial users of carbon dioxide, and damage to them would result in more carbon dioxide being released to the atmosphere.[72]

Marine Pollution and Resources

More global accords are in place to protect marine life than any other environmental problem (see Box 1.5). These accords are also some of the oldest ever agreed to by nations. The North Pacific Fur Seal Commission, established in 1911, was the first international body created to address problems of the overharvesting of marine resources. A number of fishery commissions were created in the 1930s. The International Whaling Commission (IWC) was formed in 1946 to regulate the harvesting of whales, but its decisions and guidelines were usually ignored. In 1986, worldwide concern over the rapidly diminishing whale population led to a moratorium on commercial whaling. But the commission has no enforcement powers, and a loophole in the agreement that permits whaling for "scientific purposes" has resulted in the killing of more than 14,000 whales since 1986. Much of the whaling, still permitted despite the commercial whaling ban, has been for projects of questionable scientific merit.[73] Iceland, a leading whale-hunting nation, resigned from the commission in 1992 after a proposal for harvesting whales did not satisfy the demands of its whaling industry. Much of the jurisdiction and responsibility of these international commissions has been supplanted by Exclusive Economic Zones, where quotas for fish catches are controlled by the coastal states.

BOX 1.5

PROTECTING THE WORLD'S OCEANS

■

General agreements concerning the oceans include the Convention on the High Seas, 1958; the Convention on the Continental Shelf, 1958; and the Convention on the Territorial Sea and the Contiguous Zone, 1958.

Several agreements seek to limit oil pollution:

The International Convention for the Prevention of Pollution of the Sea by Oil (1954) and the International Convention on Civil Liability for Oil Pollution Damage (CLC), adopted in 1969 and entered into force in 1975, seek to ensure that adequate compensation is available to persons affected by pollution damage through oil escaping during maritime transport by ship and to standardize international rules and procedures for determining questions of liability and adequate compensation in such areas. The 1984 protocol extended the scope into the territorial sea and the exclusive economic zones of the contracting states. There are ninety-one parties to the CLC; the secretariat is the International Maritime Organization (IMO).

The International Convention Relating to Intervention on the High Seas in Cases of Oil Pollution Casualties (Intervention Convention). Adopted at Brussels in 1969 and entered into force in 1975, the convention seeks to protect against the consequences of maritime casualties resulting in oil pollution that threatens the sea and coastline. There are sixty-seven parties to the agreement; the secretariat is the IMO.

The International Convention on the Establishment of an International Fund for Compensation for Oil Pollution Damage (Fund Convention), adopted in Brussels in 1971 and entered into force in 1978, provides for a compensation system supplementing that of the International Convention on Civil Liability for Oil Pollution Damage (CLC), to ensure full compensation to victims of oil pollution damage and to distribute the economic burden between the shipping industry and oil cargo interests. The 1984 Protocol extended the scope into the territorial sea and the exclusive economic zones of the Contracting States. There are sixty-six parties to the agreement; the secretariat is the IMO.

The International Convention on Oil Pollution Preparedness, Response, and Co-operation (OPRC) was adopted in London in 1990 and enters into force twelve months after the deposit of fifteen instruments of ratification. OPRC seeks to prevent marine pollution by oil, advance the adoption of adequate response measures in the event that oil pollution does occur, and provide for

(*Continued on page 50*)

(BOX 1.5 *Continued from page 49*)

mutual assistance and co-operation between States. There are twenty-six parties to the agreement; the secretariat is the IMO.

Three major agreements address the dumping of wastes into the oceans:

The Convention on the Prevention of Marine Pollution by Dumping of Wastes and Other Matters (London Dumping Convention), adopted in 1972 and entered into force for the United States on August 30, 1975, prohibits disposal at sea of certain wastes likely to threaten human health, harm living resources and marine life, or interfere with other legitimate uses of the sea. Permits are required to dump other wastes. Jurisdiction includes the global seas as well as territorial waters of the coastal states. Seventy-four states are parties; the secretariat is the IMO. A Protocol Relating to Intervention on the High Seas in Cases of Pollution by Substances Other than Oil was agreed to in 1973 and entered into force in 1983.

The Convention for the Prevention of Marine Pollution by Dumping from Ships and Aircraft (Oslo Convention), adopted in 1972 and entered into force in April 1974, prohibits the dumping of harmful substances into the seas by ships and aircrafts, and establishes a permitting system for dumping other materials. There are thirteen parties; the Oslo Commission serves as secretariat.

The International Convention for the Prevention of Pollution from Ships, 1973, as modified by the Protocol of 1978 (MARPOL), was adopted in 1973 in London and entered into force in 1983. It seeks to eliminate pollution of the sea by oil, chemicals, and other harmful substances that might be discharged in the course of operations; minimize the amount of oil that could be released accidentally in collisions or strandings by ships and platforms; and improve the prevention and control of marine pollution from ships, particularly oil tankers. The Convention designates the Antarctic Ocean; Mediterranean, Baltic, Red, and Black Seas; and the Persian Gulf region as special areas in which oil discharge is virtually prohibited and the wider Caribbean Sea and the North Sea as special areas subject to more stringent requirements governing the disposal of ship-generated garbage into the sea. There are ninety-five parties to the accord; thirty-five States have made exceptions for annexes III, IV, or V. NGOs participate with observer status at meetings. The secretariat is MARPOL.

The broadest agreement is the United Nations Convention on the Law of the Sea (UNCLOS), adopted in Montego Bay, Jamaica in 1982, and entered into force in November 1994. UNCLOS seeks to facilitate international communication and promote peaceful uses of the oceans, rational utilization of their resources, conservation of living resources, and the study and protection of the marine environment. It establishes principles and rules governing global and regional co-operation, technical assistance, monitoring, and environmental assessment, and adoption and enforcement of international rules and standards. It also calls for national legislation to regulate all sources of marine pollution. It is open to all states and international organizations. There are eighty-two parties and eighty-six signatories. The Secretariat is the Division for Ocean Affairs and Law of the Sea, New York.

Several regional agreements are also aimed at preventing the pollution of the sea. They usually prohibit dumping harmful substances from ships and aircraft and provide a system of permits for dumping other substances. Some provide for remedial efforts to restore damaged marine areas. Parties are limited to the states specified in the convention, but others may join following a unanimous invitation to do so from the parties.* The Convention for the Prevention of Marine Pollution from Land-based Sources (Paris Convention), adopted in June 1974 and took effect in May 1978, has served as a model for other regional agreements. The Paris convention seeks to protect the oceans and seas from land-based sources of pollution, including atmospheric pollutants as well as direct discharges into the water. It protects the waters of the northeast Atlantic, from Greenland to the Strait of Gibraltar. There are thirteen parties; the Paris Commission serves as secretariat. The 1974 and 1992 Helskinki Conventions on the Protection of the Marine Environment of the Baltic Sea Area commit parties to controlling pollution from land-based sources and from dumping that affects the Baltic Sea. There are ten parties to the 1974 agreement; six countries have signed the 1992 accord but it is not yet in force. The later convention calls on parties to apply the precautionary principle to reduce emissions that might pose a hazard, ensure that polluters pay for the pollution they produce, and to use the best environmental practices and the best available technologies.

Nine conventions are part of the United Nations' Regional Seas Programs. They generally include an action plan for cooperation on research, monitoring, pollution control and protection, rehabilitation, and development of coastal and marine resources; a legally binding framework convention embodying general principles and obligations; and detailed protocols dealing with particular

---

*The following regional agreements have been made:

Convention for the Prevention of Marine Pollution by Dumping from Ships and Aircraft (Oslo Convention), adopted in 1972; entered into force in 1974

Convention for the Prevention of Marine Pollution from Land-based sources (Paris Convention), adopted in 1974; entered into force in 1978

Convention for the Protection and Development of the Marine Environment of the Wider Carribean Region, with annex, 1983; entered into force in 1986

Convention for the Protection of the Marine Environment of the North East Atlantic (OSPAR Convention), adopted in 1992; not yet in force

Conventions for the Protection of the Marine Environment of the Baltic Sea Area (1992 Helsinki Convention), adopted in 1992; not yet in force

(Continued on page 52)

(BOX 1.5 *Continued from page 51*)
environmental problems such as oil spills, dumping, emergency cooperation, and protected areas. Participation is limited to the coastal states in the respective regions, although other states and intergovernmental integration organizations may also be invited to participate. The nine conventions are:

Convention for the Protection of the Mediterranean Sea against Pollution, 1976

Kuwait Regional Convention for Co-operation on the Protection of the Marine Environment from Pollution, 1978

Convention for the Protection of the Marine Environment and Coastal Area of the South East Pacific, 1981

Convention for Co-operation in the Protection and Development of the Marine and Coastal Environment of the West and Central African Region, 1981

Regional Convention for the Conservation of the Red Sea and Gulf of Aden Environment, 1982

Convention for the Protection and Development of the Marine Environment of the Wider Caribbean Region, 1983

Convention for the Protection, Management, and Development of the Marine and Coastal Environment of the Eastern African Region, 1985 (not yet in force)

Convention for the Protection of the Natural Resources and Environment of the South Pacific Region, 1986

Convention on the Protecting of the Black Sea against Pollution, 1992 (not yet in force)

The International Convention for the Regulation of Whaling (ICRW), adopted in Washington in 1946 and entered into force in 1948, seeks to conserve whale resources and to create an agency for the collection, analysis, and publication of scientific information related to whaling. There are thirty-nine parties. The secretariat is the International Whaling Commission (IWC), Cambridge, England.

SOURCE: Fridtjof Nansen Institute, *Green Globe Yearbook of International Co-operation on Environment and Development* (New York: Oxford University Press, 1996), 122–62.

Declining marine resources are a result of pollution and unsustainable levels of harvesting. The size of the marine catch has increased 500 percent since 1950; several fisheries, such as the Atlantic cod, haddock, Atlantic herring, Capelin, and Southern

Marine Pollution and Resources

African pilchard, have collapsed or are harvested at levels well beyond sustainable yields. The wealthy nations, after exhausting supplies near their coasts, have moved into waters near the poorer countries to increase their catch. Fish harvested by the more-developed countries are increasingly used to produce fishmeal that is fed to livestock and to manufacture fertilizer. As a result, less fish is available for direct human use; one ton of fishmeal fed to livestock produces less than one-half ton of pork or poultry. Dolphins, whales, and other marine animals have been hunted or trapped in fish nets in alarming numbers, and their survivability is uncertain.[74]

The supply of fish has been further reduced by the dumping of toxic chemicals into the oceans and the destruction of critical nurseries. Some fish have become so contaminated that humans cannot eat them safely. Ultraviolet rays from a weakened ozone layer may affect the entire ocean food chain by destroying plankton, a major source of food for ocean life. Oceans are continually absorbing vast quantities of silt and minerals carried by rivers, but it is not clear how much human-caused waste oceans can process without harming sea life. We have little knowledge of the impact on oceans of pesticides and herbicides from agricultural runoff, sewage, toxic chemicals and metals from industrial refuse, radioactive effluent, medical wastes, oil spills and refinery discharges, and other pollution. Much of this pollution initially invades estuaries, salt marshes, coral reefs, and other coastal regions, where it contaminates marine food chains. The concentration of these wastes amplifies when they contaminate higher species. Some one-third of U.S. estuarine waters are closed to shellfishing because of ecological damage. Phosphates and nitrates from sewage and fertilizer seep into oceans; as they decompose, most of the oxygen in the water is used up, choking out fish and other marine life. Dredging and construction of harbors for shipping or military bases remove coral reefs that are home to an enormous variety of marine life. Construction of cities has obliterated wetlands, formerly rich sources of sea life as well as pollution filters and buffers between sea and land.[75]

In addition, a number of international conventions to protect the oceans have been created, the most important of which are the UN

Conferences on the Law of the Sea (UNCLOS). As oil and gas reserves were discovered on the continental shelf, states began expanding their territorial limits. The 1930 League of Nations Conference initiated a half-century of conferences on the law of the sea. The First and Second UN Conferences on the Law of the Sea sought to formulate agreements to manage ocean resources. By the time the third conference was called in 1973, there was considerable support for the idea that at least the deep seabeds were a global commons and universal heritage of humankind. The third UNCLOS stretched out for nine years as delegates sought to fashion a comprehensive accord to manage fisheries, regulate harvesting of mineral resources, coordinate scientific research, ensure safe passage through international straits, impose responsibility on coastal states for protecting marine resources, and control pollution. The first countries signed the treaty in 1982, but opposition from the industrialized world, particularly the United States, has kept the treaty in limbo. The Reagan administration roundly criticized the proposal, and by 1992, only forty-nine nations had ratified it (sixty are needed for it to take effect). Despite the requirement that all provisions be accepted at ratification, several countries have tried to generate support for interim measures that conflict with the overall goals of the conference.[76] A push for support of the treaty in 1993 resulted in sufficient ratifications for it to go into effect in November 1994. The Clinton administration has sought to modify the treaty in response to business interests, but even if the treaty is modified, ratification by the Senate is uncertain.[77] As of 1997, the United States had not ratified the Law of the Sea treaty.

UNCLOS III has 439 articles that range in subject from research to pollution prevention. The most controversial provisions deal with authority to harvest the seabed. The treaty calls for the creation of an "International Sea-Bed Authority" to govern the ocean floor and for a UN body called the Enterprise. Under the agreement, miners must turn half their claims over to the Enterprise, transfer mining technology to it, and pay a $500,000 fee. The Enterprise would then share the proceeds with all countries. The treaty also provides for coastal state ownership of about 40 percent of the oceans. Coastal states are granted territorial sea rights for the first 12 nautical miles from shore, which extends their sovereign rights that distance; contiguous zone rights for 24 miles, permitting control of the seas for limited purposes; and extended economic zones for 200 miles, authorizing control over economic activity, scientific research, and environmental protection efforts. As for the continental shelf, coastal states are permitted to determine how it will be mined. The balance of the oceans is free, as it has traditionally been, but 42 percent of the seabed of this remaining free area is to be governed by the International Seabed Authority as the "common heritage of mankind."[78]

More generally, UNCLOS III became mired in conflicts between the more- and the less-developed nations. The MDCs wanted to safeguard their ability to use their technological expertise to harvest ocean-floor resources, whereas the LDCs sought to secure their rights to share in the yield. As with so many other issues, the position taken by different states and individuals depends on which reading of history is most compelling and convincing: If one believes that the LDCs have been victims of colonialism and exploitation by the wealthy states, then the MDCs have a moral obligation to ensure that this pattern does not continue. If, in contrast, one believes that states' only moral obligation is to pursue their immediate, narrow self-interest, then the MDCs ought to use their economic and political clout to ensure their own access to future resources.

UNCLOS III also provides some perspective on the viability of a comprehensive global agreement that seeks to address a wide range of issues. Since the signatories must accept the entire package, those states that object to only one provision must reject the treaty. Perhaps a more modest approach that provides for quick agreement on uncontroversial issues and recognizes the need for extended discussion of more controversial ones would be preferable to a comprehensive accord. The oceans are a complex and poorly understood set of ecosystems; current flows and marine life migration patterns do not match the way in which the oceans are carved up under UNCLOS III. Even if the convention is ratified and effectively implemented, it may be inadequate to protect the oceans. Perhaps regional agreements and plans, although they face some of these same hurdles, may be more manageable political tasks and may respect, at least to some extent, the existence of the oceans' ecosystems.

The oceans have been treated as universal waste dumps. Regulations aimed at protecting the marine environment often result in wastes simply being moved from one location to another, or transformed into another pollutant that is eventually dumped. Despite the several marine protection conventions, few provisions protect against these problems.[79] Estuaries, the zones where fresh and salt water mix, are critical elements of the environment. They are threatened by ocean and land-based pollution, but their protection is seldom provided for in international agreements.[80]

Biodiversity, Desertification, and Deforestation

The loss of biodiversity is one of the most serious problems we face.[81] Our genetic library is being depleted rapidly. At least 140 plant and animal species are condemned to extinction each day.[82] E. O. Wilson has warned that

> the worst thing that can happen . . . is not energy depletion, economic collapse, limited nuclear war, or conquest by a totalitarian government. As terrible as these catastrophes would be for us, they can be repaired within a few generations. The one process ongoing in the 1980s that will take millions of years to correct is the loss of genetic and species diversity by the destruction of natural habitats. This is the folly that our descendants are least likely to forgive us.[83]

Tropical rain forests support native peoples, serve as watersheds, absorb carbon dioxide and release oxygen, moderate air temperatures, recycle wastes, limit soil erosion, control stream and river flow and resultant flooding, and produce nuts, spices, fruits, vegetables, herbs, waxes, oils, timber, and medicines.[84] Deforestation threatens all these critical resources; it also contributes to the

Biodiversity, Desertification, and Deforestation

global climate change by reducing the absorption of CO_2 provided by forests and by increasing CO_2 emissions as plant matter decays. Humans practice deforestation to increase short-term agricultural or forest yield, create grazing lands, and to provide clearance for new human settlements. Deforestation and the loss of animal habitats translate into the extinction of perhaps 17,500 plant and animal species each year.[85]

The clearing of tropical forests for croplands may offer some immediate benefits in increasing the supply of arable land, but the long-term consequences are profound. Deforested lands quickly become nutrient-poor: In tropical forests, the vast majority of the nutrients are found in the biomass, rather than in the soil, so clearing removes most of the nutrients, and the rest leach out through rainfall. Within a few years, an ecosystem teeming with life can be reduced to sun-baked clay, with increasing erosion that makes recovery difficult and may adversely affect rivers that receive the runoff. Forests harbor great varieties of species and are the major repositories of the earth's genetic diversity.

Although much attention has been directed to the destruction of tropical rain forests, the reduction in the boreal forests of the far north that comprise nearly one-third of the world's forests is also a serious environmental threat. These forests are a major carbon sink as well as home to plants and animals and to a million indigenous people who have lived in the forests for centuries. Massive logging in Canada disrupts ecosystems, displaces native peoples, pollutes rivers with toxic chemicals used in the bleaching of pulp, and subsidizes the expenses of the large corporations such as Mitsubishi and Daishowa that have purchased leases. The subsidies are given to companies in exchange for jobs in logging and pulp mills, but the timber companies are reaping enormous profits at the expense of local residents. In Alberta, for example, the province collects $0.90 for 16 aspen trees that are made into $590 worth of pulp; a foreign mill then transforms that into $1,250 worth of paper. Taxpayers end up paying some $176,500 in subsidies for each job created at Mitsubishi's Alberta-Pacific mill. About 90 percent of the logging is clear cutting, and 25 percent of these areas do not regenerate as topsoil erodes away during and after logging. Logging also destroys permafrost, the layer below the topsoil that acts as a heat reservoir in the winter, making it more difficult for vegetation to grow. As the permafrost retreats, risks to the forests from fires and pests grows; the downward spiral contributes to the threat of global climate change as the carbon sink or absorbtion capacity is lost and carbon from dead trees is increasingly released into the atmosphere.[86]

Fertile soil is a rich, living system, comprised of millions of bacteria, yeast cells, invertebrates, and fungi. Soil is the product of an ecological process: Rocks are weathered into tiny fragments, mixed with decaying plants and animals, and moistened by rainfall and runoff. The process is so slow that, for human purposes, soil is a nonrenewable resource. Soils are fragile, highly dependent on their surrounding ecosystems for their survival—roots of grass protect soils in grassland, and leaf drops protect the soils of temperate forests, for example. Like everything else in nature, soil is part of an interconnected, interdependent relationship, connected with other environments through feedback loops and linkages. If one element is altered, the entire system may be disrupted.[87]

Droughts that dry soils, clearing of new lands to produce cash crops that earn foreign exchange to meet debt payments, saltation of land due to poorly draining fields and heavy use of fertilizers, lack of crop rotation, overgrazing, and other developments have led to loss of topsoil and soil degradation. The use of heavy farm machinery compacts soils and damages the intricate structure of microorganisms and small invertebrates that help make the soil fertile. Marginal lands are increasingly cultivated as more productive lands are less and less available to subsistence farmers.[88] Loss of topsoil and soil degradation pose a major threat to future generations: If the amount of land involved in agricultural production remains constant over the next decade, yields will have to double in order to feed the global family. Conserving healthy soil is much less expensive than repairing damaged soil, and agricultural policy efforts need to include incentives to conserve topsoil.[89]

The 1992 United Nations Convention on Biological Diversity is the latest in a long line of treaties aimed at protecting threatened

Biodiversity, Desertification, and Deforestation

species and their habitats (see Box 1.6). Negotiations began in the late 1980s under the auspices of UNEP. An ad hoc committee started work on a draft treaty in 1990; five meetings later, in May 1992, the group, which had become an "Intergovernmental Negotiating Committee," fashioned the final text of the agreement in Kenya. The agreement was an important product of the 1992 UNCED meeting, despite U.S. opposition to the proposed treaty.

The United Nations Convention on Biological Diversity had two primary goals: (1) to conserve biodiversity, defined as the "variability among living organisms from all sources . . . and the ecological complexes of which they are part;" and (2) to ensure the "fair and equitable sharing of the benefits arising out of the utilization of genetic resources."[90] The parties agreed to develop plans to protect biological diversity and to integrate these efforts with other policies.

The most controversial proposal dealt with the rights of states over their own natural resources and their development by external powers. Parties agreed that "the authority to determine access to genetic resources rests with the national governments." Access to these resources "shall be on mutually agreed terms and . . . subject to prior informed consent of the Contracting Party providing such resources, unless otherwise determined by that Party." Parties agreed to transfer "technologies that are relevant to the conservation and sustainable use of biological diversity . . . under fair and most favourable terms, including on concessional and preferential terms." Protection was also given to patents and other intellectual property rights.[91] Parties agreed to facilitate the exchange of information, promote technical and scientific cooperation, and ensure that all parties share in the benefits of research and product development.[92]

Finally, each party promised to

> take all practicable measures to promote and advance priority access on a fair and equitable basis by Contracting Parties, especially developing countries, to the results and benefits arising from biotechnologies based upon genetic resources provided by those Contracting Parties.[93]

BOX 1.6

AGREEMENTS TO CONSERVE BIODIVERSITY

■

Convention for the Preservation and Protection of Fur Seals, 1911.

Washington Convention on Nature Protection and Wild Life Preservation in the Western Hemisphere—Organization of American States, 1940.

International Convention for the Regulation of Whaling—International Whaling Commission, 1946.

Paris International Convention for the Protection of Birds, 1950.

International Plant Protection Convention, 1951.

The Antarctic Treaty, adopted in Washington D.C. in 1959; entered into force in 1961. The treaty provides that Antarctica is to be used for peaceful purposes only and ensures freedom of scientific investigation and international cooperation in Antarctica. There are forty-two contracting parties (twenty-six consultative and sixteen nonconsultative parties). Any state may join the treaty with the consent of all signatories. There is no permanent secretariat; annual meetings are hosted in rotation by the parties. The Antarctic Treaty System includes several agreements:

> The International Convention for the Conservation of Atlantic Tunas (ICCAT), adopted in Rio de Janeiro in 1966 and entered into force in 1969, seeks to maintain the population of tunas and tuna like species found in the Atlantic Ocean and the adjacent seas at levels that will permit the maximum sustainable catch for food and other purposes. The convention applies to all waters of the Atlantic Ocean and adjacent seas, including the Mediterranean Sea. There are twenty-one parties. The secretariat is the International Commission for the Conservation of Atlantic Tunas (ICCAT), Madrid.
>
> The Convention on the Conservation of Antarctic Marine Living Resources, adopted in Canberra in 1980 and entered into force in 1982, seeks to protect the ecosystem of the seas surrounding Antarctica and to conserve Antarctic marine living resources. There are twenty-nine parties, including the European Community. The Secretariat is CCAMLR, Tasmania, Australia.
>
> Two agreements—the Convention on the Regulation of Antarctic Marine Living Resources, adopted in 1988, and the Protocol on Environmental Protection to the Antarctic Treaty, adopted in 1991—are not yet in force because the requisite number of states have not ratified them.

Convention Concerning the Protection of the World Cultural and Natural Heritage (World Heritage Convention), adopted in Paris in 1972 and entered into force in 1975. The World Heritage Convention established a system of collective protection of the cultural and natural heritage of outstanding universal value and provided for emergency and long-term protection for monuments, monumental sculpture and painting, groups of buildings, archaeological sites, natural features, and habitats of animals and plants. There are 143 parties; the secretariat is the World Heritage Centre, UNESCO, Paris.

Convention on Wetlands of International Importance especially as Waterfowl Habitat (Ramsar Convention), adopted in 1971 in Ramsar, Iran and entered into force in 1975. The Ramsar Convention seeks to limit the loss of wetlands. There are ninety parties and five signatories to the convention. The secretariat is the Ramsar Convention Bureau, Gland, Switzerland.

Convention on International Trade in Endangered Species of Wild Fauna and Flora (CITES), adopted in 1973 in Washington, D.C. and entered into force in 1975. CITES places limits on international trade in threatened and endangered species of wild fauna and flora. There are 130 parties and four signatories. The secretariat is the UNEP/CITES secretariat in Geneva.

Convention on the Conservation of Migratory Species of Wild Animals (CMS), adopted in Bonn in 1979 and entered into force in 1983. The convention seeks to conserve species of wild animals that migrate across or outside national boundaries by restricting harvests, conserving habitat, and controlling other adverse factors. There are forty-three parties, including the European Community, and ten signatories. The Secretariat is the UNEP/CMS Secretariat in Bonn.

UN Food and Agricultural Organization (FAO) International Undertaking on Plant Genetic Resources, adopted in Rome in 1983 and entered into force in 1984. The agreement seeks to ensure that plant genetic resources are conserved and are accessible for plant breeding, for the benefit of present and future generations; it protects cultivated varieties of plants, plants or varieties that have been in cultivation in the past, primitive versions of cultivated plants, wild relatives of such plants, and certain special genetic stocks. Some 110 countries had signed the undertaking. The secretariat of the FAO Intergovernmental Commission on PGR acts as secretariat.

International Tropical Timber Agreement (ITTA), adopted in 1983 in Geneva and entered into force in 1985. The agreement seeks to ensure cooperation and consultation between countries that produce and consume tropical timber; promote the expansion and diversification of international trade in tropical timber; promote and support research and development to improve sustainable

(Continued on page 62)

(BOX 1.6 *Continued from page 61*)

forest management and wood utilization; encourage national policies aimed at sustainable utilization and conservation of tropical forests and their genetic resources; and ensure that all tropical timber entering international trade will be produced from forests under sustainable management by the year 2000. There are fifty-three members (twenty-three producing and twenty-seven consuming members), including the European Community. The secretariat is the International Tropical Timber Organization International Organizations Center, Nishi-ku Yokohama, Japan.

Convention on Biological Diversity, adopted in Rio de Janeiro in 1992 and entered into force in 1993. The convention seeks to ensure conservation of biological diversity, promote a fair and equitable sharing of the benefits arising out of the commercial use of genetic resources, and provide for the transfer of relevant technologies and resources to less developed nations. There are 135 parties, including the European Community, and forty-two signatories. The secretariat is the UNEP/Interim Secretariat for the Convention on Biological Diversity, Geneva, Switzerland.

The Convention to Combat Desertification (CCD), adopted in Paris in 1994, has not yet taken effect. As of November 1995, there were thirteen parties and 102 signatories. The CCD seeks to develop cooperative efforts among donors, NGOs, national governments, and international organizations that are consistent with sustainable development to combat desertification and mitigate the effects of drought and desertification. Separate annexes address specific efforts aimed at Africa, Latin America, Asia, and the Caribbean.

SOURCE: Fridtjof Nansen Institute, *Green Globe Yearbook of International Co-operation on Environment and Development* (New York: Oxford University Press, 1996), 163–84.

The parties agreed to establish and support a mechanism for channeling financial resources to the developing countries that became part of the Global Environment Facility.[94] The agreement established a dispute resolution process to arbitrate disputes that may arise under the treaty.

The Convention entered into force on December 29, 1993; as of 1995, 165 countries had signed the Convention and forty-two had ratified it. The Intergovernmental Committee on the Convention on Biological Diversity, created by UNEP under authority of the Convention, assists parties in complying with their commitments under the agreement.[95] Agreements to protect biodiversity are

Population

among the oldest international environmental accords in place, and have been aimed at specific species as well as at geographic regions.

Population

The environmental impact of human activity is a function of (1) population and consumption, and (2) the vulnerability of the ecosystem. There are feedback loops from social and environmental effects; both social and ecological factors are important, including patterns of land ownership, distribution of wealth, community structures, economic activity, political and economic stability, and psychological views of the interaction of humans and nature. Uncertainty about population growth, energy consumption, economic activity, and natural resource conditions pervades these determinations. Ecological systems can withstand external shocks and influences until they are no longer resilient and even small increments can precipitate dramatic change. Population is often a key force in such change. Many assume that the developing world will follow the same demographic pattern the developed countries experienced: a declining death rate, followed by a declining birth rate, resulting from increased wealth and social change that empowers women to choose fewer children. But without continual economic and social progress, population growth may not be stabilized.[96]

The human family included two and a half billion people in 1950. Some 37 million children were born that year. Forty years later, the world family had doubled in size, and each year more than 90 million new residents of the planet, the equivalent of repopulating Mexico, are born. More than 90 percent of population growth occurs in the less-developed world, where the fertility rate is just below four childbirths per woman; in the wealthy world, the rate is about two.[97]

The steady fall of the fertility rate gives the misleading impression that population pressures are abating. However, the number of women of childbearing age has ensured that population growth continues. Throughout the 1960s and 1970s, the global population growth rate was 2.0 percent; by 1990 it had fallen to 1.6 percent. But population growth in some regions continues to place great pressures on the biosphere.[98] At current growth rates, the population will double again in the next thirty-five or forty years. Even if fertility rates decrease, the number of births will almost certainly continue to increase, because the number of women of childbearing age is expected to grow from about 1 billion in 1994 to 1.5 billion in 2010. And even if those women have, on average, only two children, world population will continue to expand. Population growth seems inevitable, unless death rates dramatically increase.[99]

Some projections of world population growth forecast that the population will grow to about 8 billion within the next twenty to thirty years; projections beyond that grow increasingly uncertain, since behavior and fertility are so difficult to estimate for longer periods of time. After 2010–2020, population estimates diverge considerably, depending on which assumptions are used, although many studies conclude that a population of 12 to 15 billion by the years 2040–2070 is almost inevitable. The State of World Population report, issued by the United States, projected that the current population of 5.66 billion was growing by 94 million a year, and, if trends continue, would grow to 8.5 billion in 2025, despite a continued decline in overall fertility rates. But since fertility rates are so difficult to predict, population growth could also go as high as 12.5 or as low as 7.8 billion by the mid-twenty-first century. At current growth rates, the population in the less-developed world will double by the year 2020. Although some countries, such as China and Costa Rica, have been able to check their population growth, other countries, such as Mexico and Brazil, continue to grow rapidly, despite some industrialization and considerable economic growth.[100]

The challenge of slowing population growth in specific areas is daunting. In sub-Saharan Africa, for example, almost one-half of the population is under the age of fifteen. In order to stabilize population over the next thirty years, each family can have no more than one child. Latin America and Asia, with more than one-third of their population under fifteen, must have one-child families for the next twenty-five years to check population growth.[101] Since such a severe shift in behavior can hardly be envisioned, population stabilization will take much longer, unless it is overtaken by famine, war, ecological destruction, or some other adjustments made by the ecosystem. China, which succeeded in reducing its birthrate from forty-three per thousand in 1970 to eighteen per thousand in 1979, has been criticized for coercing its citizens. But some argue that "the horrors of children dying of starvation in the overpopulated and poorer parts of China over many years far outweigh the drastic powers of persuasion used to reduce births."[102]

The inexorable pressure of population growth contributes greatly to global environmental problems, but it also interacts with other trends, including the unequal consumption of energy and other resources. Natural resource problems include population growth, the inability to meet the projected demand for food, a distribution system that currently falls well short of delivering the food that is produced, the depletion of essential minerals, and reliance on energy sources that are increasingly expensive to harvest. Pollution of air, water, land, and stratosphere further complicate the problems. Population growth and harvesting of resources for human needs are expected to diminish considerably the resources that will be needed to supply future needs. The issue of population growth is also intertwined with arguments that high consumption levels in the industrialized countries are a greater threat to the global environment than is the high population growth in the developing countries.

The relentless growth of the global family has engendered a long-running debate over the need for population control.[103] About $6 billion is spent worldwide on population stabilization efforts, with $1 billion coming from Japan, the United States, and Europe, and the balance from the less-developed nations themselves. Some 450 million people in the developing world use contraceptives, although use varies considerably. In Niger, for example, only about 4 percent of couples use contraceptives. The optimistic view is that population growth is no real threat, because technological advances will permit us to accommodate more and more people. The pessimistic view is that we cannot continue to grow and consume resources at current rates, and we must make dramatic changes in our behavior or the ecosystems will no longer be able to sustain life.

Optimists argue that history demonstrates the likelihood that population growth will be stabilized at sustainable levels. Just as European countries have completed a demographic transition from an exploding birthrate that helped energize the industrial revolution to near-zero growth rates, so too will the less-developed world. Others argue that modern technology has permitted and will continue to permit us to grow and improve quality of life, and that we are indeed capable of managing natural systems. The earth is sufficiently resilient, and human ingenuity sufficiently creative, that we will survive and adapt as population grows.[104]

Optimists also argue that proponents of reducing population growth wrongly believe that rapid population growth impedes economic progress. They point to areas, such as the emerging industrial powerhouses in East Asia, where population surges and material advances seem consonant. They claim that technological advances will enable food production to keep up with growing demand. Expanding the availability of contraceptives will not help reduce population growth; markets will inevitably send the correct signals to people by encouraging development and discouraging fertility.

Pessimists argue that even if there is historical evidence of a self-correcting mechanism, we now do not have the centuries that Europeans had to make those adjustments gradually. This group, often ecologists and biologists, believe that the earth cannot absorb the consequences of such an increase in population. As developing countries industrialize, increasing levels of pollution threaten air and water quality and overall climate stability. From this view, unchecked population growth is the most pressing environmental problem.[105] Others have been warning for decades that significant limits to population growth are rooted in scarce, critical natural resources and the patterns of Western consumer-driven life.[106] Triggers in prices and markets that result from increasing scarcity will not bring back lands that have been paved over or species of crops and animals that have been lost.[107]

Population

The International Conference on Population and Development (ICPD), held in Cairo in September 1994, was the third in a series of major conferences on global population that began in Bucharest in 1974. In the 1974 meeting, representatives from the less-developed nations pressed for more development assistance from the wealthy nations. Although some participants brought up the issue of improving the status of women, the debate lay elsewhere. The United States became the largest contributor to population stabilization programs. However, at the 1984 global population meeting, the United States strongly opposed funding for any reproductive services that included abortion. The Reagan administration, in a 1984 executive order, prohibited U.S. funds from going to the International Planned Parenthood Federation because it included abortion in its list of services. In 1985, the U.S. Congress banned funding for groups involved in coerced abortions or sterilization, including the UN Population Fund. The 1984 Mexico City conference provided little opportunity for proponents of improving the status of women to participate in the proceedings. Since the official delegates were mostly men, women's participation was generally limited to nongovernmental organizations, but NGOs were not even given space to meet.[108]

The preparatory meetings for the Cairo conference established the goal of stabilizing the global population at 7.8 billion by 2050. However, the negotiations did not focus so much on this numerical goal as on ways to improve the status of women. This reflected, in part, a decade of efforts by women's rights advocates to focus attention on the wide range of women's issues—access to reproductive services, credit, health care, literacy and education programs, governing institutions, and land to grow food for their families. Despite some success in reducing population growth through family-planning services, Cairo conference participants, for the most part, accepted the position that expanding access to health care, education, and economic opportunities is a more effective approach to population control; this approach also promotes equality. Many women were optimistic because, for the first time, a global conference recognized that population growth will not be slowed until it is addressed as part of a broad set of interrelated problems including poor health care, lack of choices about family planning, abuse and violence, and general powerlessness of women.[109] Women were no longer to be, in the words of one delegate, the "receptors of contraceptives," but, rather, "agents of change."[110]

The Cairo meeting also addressed broader issues of development as a consequence of the 1992 Rio conference, which highlighted the growing tension between the more-developed and the less-developed nations. As the United Church of Christ argued in a document prepared for Cairo's preparatory conference,

> Population growth and overconsumption among the wealthy put unparalleled pressure on the earth's fragile and often irreplaceable environment, but wealth offers protection for a time against the consequences of this folly: for the poor, however, the consequences are immediate and devastating.[111]

For others, the emphasis on international family planning programs was a continuation of policies aimed at reducing the number of children of color born into the world. Underlying the Cairo conference, critics argued, was racism: Since the "fastest population growth is among people who have black, brown or yellow skins, . . . if white folks don't do something they will be overwhelmed by the masses of a different color."[112] Critics also claimed that "agencies, controlled and funded by those of Christian/European heritage, [were] seeking to limit the number of 'brown' babies through imposition of Western concepts of family planning."[113]

Others complained that the Cairo agenda was dominated by a commitment to the "sexual revolution and to the notion of liberty as radical personal autonomy."[114] Even some family-planning activists were dismayed at efforts to push the agenda beyond integrating health care, education, female empowerment, and family planning, to promoting sexual freedom and activity among adolescents.[115]

The major controversy at the Cairo conference swirled around the impact of the agreement on abortion. Pope John Paul II expressed fear that the Cairo meeting would promote contraceptive use,

Population

abortion, sexual activity among unmarried persons, and alternative "lifestyles" in marriage. Concern focused on phrases such as "reproductive health" that were believed to imply an unrestricted right to abortion.[116] The pope's agenda appeared to be to block any endorsement of abortion in the Cairo Conference on Population and Development's Program of Action. But abortion was not the major issue in the Program of Action: The program gave much more attention to contraceptives than to abortion. Although abortion was to be provided and made safer in countries where it was legal, a major thrust of the conference was to expand dramatically the availability of contraceptives. Since both abortion and contraception are rejected by the Vatican, perhaps the greater concern was with contraceptives. However, Vatican diplomats did not insist that parentheses be placed around provisions that discussed access to contraceptives. (This use of parentheses is the diplomatic means of registering dissension or concerns with specific proposals.[117]) The Vatican eventually endorsed the Program of Action's declaration of principles, as well as chapters on the family, stimulating economic growth, migration, and empowering women. Some twenty delegates registered reservations on language dealing with sex and abortion.[118]

The pope was, in perspective, quite successful in attracting attention to his concerns, and in encouraging those who believe that population growth is not among the fundamental global threats. Some critics argued that religious leaders played a more powerful role when they pointed out the consequences of greed and materialism, vices that are bad for the environment as well as for the human spirit.[119] Some Muslim leaders lined up with the Vatican to condemn the draft agreement as a violation of Islamic law, including provisions that referred to providing teenagers with confidential sexual health care, "marriage and other unions," "sexually active unmarried individuals," and provisions that required men and women to receive equal shares of inheritances.[120] Some feared that the talks would bring the results of the "West's sexual revolution" to the Muslim world.[121]

U.S. officials tried to defuse the tension with the pope by rejecting support for abortion on demand or abortion as a method of family planning.[122] Nevertheless, the Vatican launched a very specific attack on the United States because of its role in the preparation of the agreement, which was described as "cultural imperialism" whereby "abortion on demand, sexual promiscuity, and distorted notions of the family are proclaimed as human rights."[123] Vatican officials argued that economic reform should be the major focus of global leaders: "Demographic growth is the child of poverty. . . . Rather than reduce the numbers at the world's table, you need to increase the courses and distribute them better."[124] "Life is sacred," one Vatican spokesman wrote, and "human enterprise and freedom can provide for human needs . . . . Reform of society must begin in a reform of the heart of the individual and an accurate understanding of the worth of each human life."[125] Other religious groups and foundations launched campaigns aimed at drawing attention to overconsumption in the United States and other wealthy nations, paralleling arguments of economists that the United States needs to save and invest more and consume less.[126]

The "Program of Action," a 113-page document approved in April 1994 by the UN committee responsible for the Cairo meeting, provides the basic outline for a twenty-year plan to promote women's rights and reduce population growth to 7.27 billion by 2015. The final document did not satisfy the Vatican. It provided that abortions "should be safe" where they are legal, and that they should "in no case . . . be promoted as a method of family planning."[127] A central theme embraced by nearly every one of the 150 delegates was the empowerment of women.[128] Although few participants were completely happy with the results of the report, as one observer put it, "conferences and documents like these never get out ahead of where the people who agree to them are comfortable. What these conferences do is take an aerial snapshot of where we're at." The provisions are not binding on governments and do not take precedence over national laws or religious beliefs. But the agreement can contribute to the building of consensus over the interaction of population growth, poverty, and economic development.[129] It calls for increasing spending to $17 billion for population programs, one-third to come from the MDCs. That would require the United States to increase its donations from the fiscal year 1995 level of $585 million to $1.9 billion by the year 2000.[130] Japan's contribution was projected to reach $1.2 billion by the same date.[131] Spending was to include about $10.2 billion for family planning and $5 billion for reproductive services.

One of the most promising products of the conference was the creation of "Partners in Population and Development: A South-South Initiative," announced by ten developing countries that came together to share their successes in reducing population growth. These successful strategies include offering a wide range of choices for family planning and careful integration with local cultural and political conditions.[132]

Many challenges remain, however, in determining how to finance and deliver family planning and related services in ways that are consistent with local religious, cultural, social, and economic conditions. In many less-developed countries, family-planning services are still widely seen as a first-world idea. Much of the focus will have to be on those countries that have the greatest population growth rates and the fewest resources to help empower women.[133]

Other Agreements

The Convention on Environmental Impact Assessment in a Transboundary Context (Espoo Convention), adopted in 1991, seeks to encourage international cooperation in evaluating environmental impacts and preventing significant adverse impacts, especially those that cross national boundaries. It requires notification and consultation among states and encourages public participation for major projects under consideration that are likely to cause significant adverse environmental impact across boundaries. It also encourages the development of consistent national policies and practices for preventing, mitigating, and monitoring significant adverse environmental impacts. As of 1995, eleven countries were parties and twenty were signatories.

The Convention on the Ban of the Import into Africa and the Control of Transboundary Movements and Management of Hazardous Wastes within Africa was adopted in 1991 in Bamako.

The accord seeks to protect human health and the environment from dangers posed by hazardous wastes by reducing generation of these wastes to a minimum in terms of quantity and harmful potential. Participation is limited to member states of the Organization of African Unity (OAU). As of 1995, there were nine parties and seventeen signatories. As a result of the agreement, a large portion of the world has been closed off to the hazardous waste trade. The bulk of hazardous waste imports has shifted from Africa to the Caribbean, South and Central America, and Asia. Illegal activities persist, and countries seek loopholes so that the trade can continue[134]. In 1997, North Korea announced it had agreed to accept shipments of nuclear wastes from Taiwan, raising new fears about international trade in hazardous materials. The agreement promised to help raise cash for impoverished North Korea and solve Taiwan's waste storage problem, but South Koreans protested the proposal because storage of the waste would take place less than 40 miles from the disposal site and because they feared the North Koreans were unprepared to store the wastes safely.[135]

Assessing International Environmental Agreements

Global agreements have been difficult to put in place. Many important agreements, such as those that address compensation for injury from nuclear power plants and those that seek to protect the seabed as a global heritage, have simply not been ratified by major nations. Although the first Law of the Sea conference was held in 1973, for example, not until 1982 was a draft treaty formulated. The 1973 International Convention for the Prevention of Pollution from Ships placed modest limits on ocean dumping, but it took the United States fourteen years to ratify, and dumping of plastics by the military was not limited under the treaty until 1994. The 1972 London Dumping Convention prohibits ocean disposal of heavy metals, specific carcinogens, and radioactive wastes, but this hasn't really been enforced. Loopholes in these agreements are legendary. But some progress has been made in regulating the shipping of

Assessing International Environmental Agreements

hazardous wastes; in addressing the long-range transportation of air pollutants, primarily acid rain; and in protecting the ozone layer. Much of the criticism has focused on the national and international institutions responsible for administering international environmental laws.[136]

Global environmental accords have varied greatly in their effectiveness. Some have resulted in multilateral research and monitoring activities that have made significant contributions to knowledge of global environmental problems. Many of them, however, lack basic enforcement mechanisms or sanctions or are unenforceable. In domestic law, legal provisions are binding and enforcement mechanisms are already in place. However, nations are free to ignore global agreements. Countries will generally comply with treaties when provisions are mutually advantageous or when issues are narrowly focused and relatively insignificant; however, international law puts few constraints on nations that pursue international power and self-interest.[137] The agreements usually do not permit aggrieved parties to use international adjudicatory processes to pursue claims of violation. They mainly establish reciprocal procedures parties can use against those who breach the agreement, and they also rely heavily on pressure from other parties, domestic interests, and publicity from nongovernmental organizations to encourage compliance.[138]

International environmental agreements have produced some notable accomplishments, but progress so far is dwarfed by the remaining tasks. Information on trends in environmental quality and natural resource preservation is still sketchy. Many other problems lack global attention. Many people believe that we have failed to meet much of the challenge posed by the deteriorating global environment. American institutions have, for the most part, been unable to address effectively the difficult issues confronting us. The lack of political leadership has been matched by citizens' lack of willingness to make the kinds of sacrifices necessary to begin to solve these problems. Other countries have had more success than the United States in reducing some greenhouse gas emissions, for example, and in consuming less energy. But these differences include geographic and other factors as well as variations in government and politics. Because Japan and Europe face more severe resource constraints than the United States, they have already learned to become more energy efficient and to rely more on mass transit. Japan, perhaps more than any other industrialized nation, has been able to reduce air and water pollution and waste disposal.

The ideology of growth, the reliance on technology, and the primacy of economic concerns have produced similar outcomes in the East and the West. The countries of Eastern Europe provide striking examples of the consequences of environmental neglect and demonstrate that these environmental ills transcend types of political regimes. The lack of openness and public scrutiny of government efforts exacerbated the problems as government production quotas resulted in tremendous pressures to externalize pollution and injuries to the environment.[139]

Global environmental agreements represent important statements concerning emerging global expectations. Despite their shortcomings, they provide a framework on which future agreements can be built. The countries of the world face two primary tasks. First, implementing these agreements will take an enormous commitment of effort, resources, and cooperation. Second, even as the provisions of these agreements are implemented, countries will be involved in a variety of negotiations to strengthen them and formulate accords to solve other problems. Ensuring that the goals of these international environmental agreements are achieved entails many difficult challenges. These challenges are the subject of the chapters that follow.

References

1. Tony Brenton, *The Greening of Machiavelli: The Evolution of International Environmental Politics* (London: Earthscan, 1994), 36.
2. Erik P. Eckholm, *Down to Earth: Environment and Human Needs* (New York: Norton, 1982), 3–5.
3. William K. Stevens, "Earth Summit Finds The Years of Optimism Are A Fading Memory," *New York Times*, 9 June 1992, B10.

References

4. See John McCormick, *Reclaiming Paradise: The Global Environmental Movement* (Bloomington: Indiana University Press, 1989), 90–104.
5. Barbara Ward and René Dubos, *Only One Earth* (Harmondsworth, Middx.: Penguin, 1972).
6. United Nations Conference on the Human Environment, *Stockholm Declaration of the United Nations Conference on the Human Environment* (16 June 1972) UN Doc. A/CONF. 48/14/Rev. 1 at 3 (1975).
7. For a review of the UNCED meeting see Michael Keating, *Agenda for Change: A plain-Language Version of Agenda 21 and the Other Rio Agreements* (Geneva: Center for Our Common Future, 1994); Michael Grubb, et al., *The Earth Summit Agreements: A Guide and Assessment* (London: Earthscan, 1993); Adam Rogers, *The Earth Summit: A Planetary Reckoning* (Los Angeles: Global View Press, 1993); Daniel Sitarz, *Agenda 21: The Earth Summit Strategy to Save Our Planet* (Boulder, Colo.: Earthpress, 1993); United Nations, *Agenda 21: Programme of Action for Sustainable Development; Rio Declaration on Environment and Development; Statement of Forest Principles* (New York: United Nations, 1992).
8. "PrepCom 4 Adopts Draft Declaration of Principles for Rio Summit," *Earth Summit Update* (April 1992): 2.
9. *Rio Declaration on Environment and Development*, Principle 2.
10. "PrepCom 4 Forwards Agenda 21 to Rio; Many Ambitious Proposals Are Blocked," *Earth Summit Update* (April 1992): 1, 5; "Agenda 21 Chapter-by-Chapter Summary," *Earth Summit Update* (May 1992): 4–5.
11. "Pressure For Specific Funding Level Expected in Rio," *Earth Summit Update* (May 1992): 2.
12. Frank Edward Allen and Rose Gutfeld, "Earth Summit Neglects Major Issues Of Poverty, Drought and Population," *Wall Street Journal*, 11 June 1992, 1.
13. Lee A. Kimball, *Forging International Agreement* (Washington, D.C.: World Resources Institute, 1992), 75–76.
14. Barbara Crossette, "UN Endorses a Treaty to Halt All Nuclear Testing," *New York Times*, 11 September 1996.
15. See Marvin S. Soroos, "The Partial Test Ban Treaty of 1963: A Case Study of A Successful International Environmental Regime" (paper presented at the 36th annual convention of the International Studies Association, Chicago, 21–25 February, 1995).
16. "Approve the Test-Ban Treaty," *New York Times*, 10 September 1996, A20.
17. India, Libya, and Bhutan voted against the treaty; five countries abstained: Cuba, Lebanon, Mauritius, Syria, and Tanzania; nineteen countries were absent or were ineligible to vote because their UN

assessments are in arrears. See Barbara Crossette, "UN Endorses a Treaty to Halt All Nuclear Testing."

18. The United States, Russia, China, Great Britain, and France.
19. World Resources Institute, *World Resources 1992–93: A Guide to the Global Environment* (Washington, D.C.: World Resources Institute, 1992), 197–99.
20. Cecie Starr and Ralph Taggart, *Biology*, 5th ed. (Belmont, Calif.: Wadsworth, 1989), 795.
21. The acidity of a substance is indicated by its pH factor, measured on a logarithmic scale. A change of one unit on the scale represents a tenfold increase or decrease in acidity. Battery acid has a pH rating of 1, lemon juice's pH rating is approximately 2, and pure water is rated at 7, the neutral position. Natural sources cause rainfall that is otherwise unpolluted to have a pH factor of 5.5 to 7.0. Rain rated below 5.6 is generally considered to be sufficiently acidic to have negative environmental consequences.
22. Library of Congress, Congressional Research Service, "Environmental Protection Laws and Treaties: Reference Guide" (30 January 1991), 33–35. See also Environmental Protection Agency, *U.S. Government Participation in International Treaties, Agreements, Organizations and Programs* (1984), A–2; and Department of State, *Treaties in Force: A List of Treaties and Other International Agreements of the United States in Force on January 1, 1988* (Washington, D.C., 1988), 322–23.
23. For a review of these developments, see C. Ian Jackson, "A Tenth Anniversary Review of the ECE Convention on Long-Range Transboundary Air Pollution," *International Environmental Affairs* 2, no. 3 (1990): 222–25.
24. Hilary French, *After the Earth Summit: The Future of Environmental Governance*, Worldwatch Paper 107 (Washington, D.C.: Worldwatch Institute, March 1992), 10–11.
25. Canadian Environmental Law Research Foundation (Toronto) and the Environmental Law Institute (Washington, D.C.), *The Regulation of Toxic and Oxidant Air Pollution in North America* (Toronto: CCH Canadian Limited, 1986), 177–79.
26. Interview with Drew Lewis, *EPA Journal* (June–July 1986): 4–7.
27. American Enterprise Institute, *The Clean Air Act: Proposals for Revision* (Washington, D.C.: AEI, 1981), 76–78.
28. The acid rain title of the 1990 amendments to the Clean Air Act included several major provisions:

> Sulfur dioxide emissions, the major precursor of acid rain, will be cut in half by imposing emission limitations on utility power

References

plants. Sulfur dioxide emissions will be cut by 10 million tons annually from 1980 levels by the year 2000.

Nitrogen oxide emissions, the other major cause of acid rain, will be reduced by 2 million tons from 1980 levels.

The EPA will allocate allowances or permits for each source of sulfur dioxide. An allowance will be given for each ton of emission permitted; sources cannot release emissions beyond the number of allowances they have. Allowances may be traded, bought, or sold among allowance holders.

Additional allowances will be given to certain Midwest utilities that they can sell in order to help finance their cleanup efforts. The EPA is also required to create an additional pool of allowances to permit construction of new sources or expansion of existing ones in states with cleaner air.

(*Clean Air Act Amendments of 1990*, section 404)

29. For a discussion of the evolution of acid rain provisions in general and the emissions trading system in particular, see the following hearings held in the House of Representatives that addressed acid rain: Committee on Energy and Commerce, Subcommittee on Energy and Power, *Acid Rain Oversight*, Serial No. 100-222 (1989); Committee on Energy and Commerce, Subcommittee on Energy and Power, *Clean Coal Technologies*, Serial No. 100-70 (1988), Committee on Energy and Commerce, Subcommittee on Energy Conservation and Power, *Acid Deposition Control Act*, Serial No. 99-153 (1987); Committee on Energy and Commerce, Subcommittee on Fossil and Synthetic Fuels, *Clean Coal Technologies—Part 2*, Serial No. 99-111 (1986); Committee on Energy and Commerce, Subcommittee on Fossil and Synthetic Fuels, *Future of Coal*, Serial No. 98-146 (1984); Committee on Energy and Commerce, Subcommittee on Health and the Environment, *Acid Deposition Control Act of 1986 (Part 1)*, Serial No. 99-85 (1986); Committee on Energy and Commerce, Subcommittee on Health and the Environment, *Acid Deposition Control Act of 1986 (Part 2)*, Serial No. 99-86 (1986); Committee on Energy and Commerce, Subcommittee on Health and the Environment, *Acid Deposition Control Act of 1986 (Part 3)*, Serial No. 99-87 (1986), Committee on Energy and Commerce, Subcommittee on Health and the Environment, *Acid Deposition Control Act of 1987*, Serial No. 100-96 (1988); Committee on Energy and Commerce, Subcommittee on Health and the Environment, *Acid Rain Control Proposals*, Serial No. 101-25 (1989); Committee on

Energy and Commerce, Subcommittee on Health and the Environment, *Acid Rain in the West*, Serial No. 99-49 (1986).
30. See, for example, Government of Canada, "Economic Instruments for Environmental Protection," discussion paper (1992).
31. See generally, Richard Elliot Benedick, *Ozone Diplomacy: New Directions in Safeguarding the Planet* (Cambridge, Mass.: Harvard University Press, 1991); David E. Fisher, *Fire and Ice: The Greenhouse Effect, Ozone Depletion and Nuclear Winter* (New York: Harper & Row, 1990); John Gribbin, *The Hole in the Sky* (New York: Bantam Books, 1988); S. Fred Singer, *Global Climate Change: Human and Natural Influences* (New York: Paragon House, 1989); and Karen T. Liftin, *Ozone Discourses: Science and Politics in Global Environmental Cooperation* (New York: Columbia University Press, 1994).
32. See, generally, Cynthia Pollock Shea, "Protecting Life on Earth: Steps to Save the Ozone Layer," Worldwatch Paper 87 (Washington, D.C.: Worldwatch Institute, 1988); Richard S. Stolarski, "The Antarctic Ozone Hole," *Scientific American* 258 (January 1988): 35; "Holes Galore," *The Economist* (15 October 1994): 114; and Karen Litfin, *Ozone Discourses* (New York: Columbia University Press, 1994).
33. Peter H. Sand, "The Vienna Convention Is Adopted," *Environment* 27 (June 1985): 20.
34. Ibid.
35. David D. Doniger, "Global Emergency," *Environmental Forum* (July/August 1988): 17.
36. The Soviet Union told the meeting that it would abide by the Montréal Protocol, but that it would not support a drive led by the United States and the European Community for a total ban on CFCs by the turn of the century.
37. Don Hinrichsen, "Stratospheric Maintenance: Fixing the Ozone Hole is a Work in Progress," *Amicus Journal* (Fall 1996): 35–38.
38. Mike Mills, "Ratification of Ozone Pact Recommended," *Congressional Quarterly Weekly Report* 46 (20 February 1988): 370.
39. Mostafa K. Tolba, "The Ozone Agreement—and Beyond," *Environmental Conservation* 14, no. 4 (1987): 290.
40. Richard Elliot Benedick, *Ozone Diplomacy*, 8–9.
41. Associated Press, "Data Point to Ultimate Closing of Ozone Hole," *New York Times*, 31 May 1996, A11.
42. Thomas F. Homer-Dixon, "Physical Dimensions of Global Change," in, *Global Accord: Environmental Challenges and International Responses* ed. Nazli Choucri(Cambridge, Mass.: MIT Press, 1993), 43–66.
43. See Eugene Lindon, "Who Lost the Ozone?" *Time*, 10 May 1993, 56–57.

References

44. See Dean Edwin Abrahamson, ed., *The Challenge of Global Warming* (Washington, D.C.: Island Press, 1989); Stewart Boyle and John Ardill, *The Greenhouse Effect* (Kent, England: New English Library, 1989); Gary C. Bryner, ed., *Global Warming and the Challenge of International Cooperation: An Interdisciplinary Assessment* (Provo, Utah.: David M. Kennedy Center, Brigham Young University, 1992); John Erickson, *Greenhouse Earth: Tomorrow's Disaster Today* (Blue Ridge Summit, Pa.: TAB Books, 1990); Jim Falk and Andrew Brownlow, *The Greenhouse Challenge: What's To Be Done?* (New York: Penguin, 1989); David L. Feldman, ed., *Global Climate Change and Public Policy* (Chicago: Nelson Hall, 1994); David E. Fisher, *Fire and Ice: The Greenhouse Effect, Ozone Depletion and Nuclear Winter* (New York: Harper & Row, 1990); John Gribbin, *Hothouse Earth: The Greenhouse Effect and Gaia* (New York: Grove Weidenfeld, 1990); John Gribbin, *The Breathing Planet* (Oxford: Basil Blackwell, 1986); Sherwood B. Isdo, *Carbon Dioxide and Global Change: Earth in Transition* (Tempe, Ariz.: IBR Press, 1989); Jeremy Leggett, *Global Warming: The Greenpeace Report* (Oxford: Oxford University Press, 1990): Henry Lee, *Shaping National Responses to Climate Change: A Post-Rio Guide* (Washington, D.C.: Island Press, 1995); Francesca Lyman et al., *The Greenhouse Trap: What We're Doing to the Atmosphere and How We Can Slow Global Warming* (Boston: Beacon Press, 1990); Terrell J. Minger, ed., *Greenhouse Glasnost: The Crisis of Global Warming* (New York: Ecco Press, 1990); Natural Resources Defense Council, *Cooling the Greenhouse: Vital First Steps to Combat Global Warming* (Washington, D.C.: NRDC, 1989); Michael Oppenheimer and Robert H. Boyle, *Dead Heat: The Race Against the Greenhouse Effect* (New York: Basic Books, 1990); Fred Pearce, *Turning Up the Heat: Our Perilous Future in the Global Greenhouse* (London: Paladin Grafton, 1989); Norman J. Rosenberg et al., *Greenhouse Warming: Abatement and Adaptation* (Washington, D.C.: Resources for the Future, 1989); Stephen H. Schneider, *Global Warming: Are We Entering the Greenhouse Century?* (San Francisco: Sierra Club Books, 1989); Mark C. Trexler, *Minding the Carbon Store: Weighing U.S. Forestry Strategies to Slow Global Warming* (Washington, D.C.: World Resources Institute, 1991); Donald J. Wuebbles and Jae Edmonds, *Primer on Greenhouse Gases* (Chelsea, Mich.: Lewis Publishers, 1991).

Government studies and reports include the following: U.S. Congress, Office of Technology Assessment, *Climate Treaties and Models: Issues in the International Management of Climate Change* (Washington, D.C.: GPO, 1994); U.S. Congress, Office of Technology Assessment, *Preparing for an Uncertain Climate* (Washington, D.C.: GPO, 1993); OTA, *Changing by Degrees: Steps to Reduce Green-*

house Gases (Washington, D.C.: GPO, 1991); Environmental Protection Agency, *The Potential Effects Of Global Climate Change On The United States* (Washington, D.C.: EPA, 1989); EPA, *Policy Options for Stabilizing Global Climate* (Washington, D.C.: EPA, 1990); Department of Energy, Energy Information Agency *Emissions of Greenhouse Gases in the United States 1985–1990* (Washington, D.C.: Department of Energy, 1993).

45. Irving M. Mintzner, *A Matter of Degrees: The Potential for Controlling the Greenhouse Effect* (Washington, D.C.: World Resources Institute, 1987), 5; Natural Resources Defense Council, "Cooling the Greenhouse Effect: Vital First Steps To Combat Global Warming" (Washington, D.C.: NRDC, May 1989), 11–12.
46. Barry Commoner, *Making Peace with the Planet* (New York: Pantheon, 1990), 6.
47. Christopher Flavin and Odil Tunali, "Getting Warmer: Looking for A Way Out of the Climate Impasse," *World-Watch* 8 (March/April 1995): 10.
48. Natural Resources Defense Council, "Cooling the Greenhouse: Vital First Steps to Combat Global Warming."
49. Interparliamentary Conference on the Global Environment, "Legislative Strategy Papers" (April 29–May 2, 1990).
50. "Climate Institute Develops Draft Global Warming Framework Treaty," *Greenhouse Effect Report* (March 1990): 18.
51. United Nations Framework Convention on Climate Change, Article 2.
52. Ibid., Article 4, section 1.
53. Ibid., Article 4, section 2 a,b; section 3.
54. Ibid., Articles 7, 8, 9, 10, and 12.
55. Fridtjof Nansen Institute, *Green Globe Yearbook of International Cooperation on Environment and Development* (New York: Oxford University Press, 1995), 117.
56. Christopher Flavin and Odil Tunali, "Getting Warmer: Looking For A Way Out of the Climate Impasse," *World-Watch* 8 (March/April 1995): 10–19, at 12, 15.
57. Department of State, *National Action Plan for Global Climate Change* (Washington, D.C.: 1992); President William J. Clinton and Vice President Albert Gore, Jr., *The Climate Change Action Plan* (October 1993); Department of Energy, *The Climate Change Action Plan: Technical Supplement* (Springfield, Va.: National Technical Information Service, 1994).
58. President William J. Clinton and Vice President Albert Gore, Jr., *The Climate Change Action Plan* (October 1993).
59. Mobil Oil advertisement, *New York Times*, 8 June 1995.

References

60. Kent Jeffreys, "More Hot Air on Greenhouse Gases," *Wall Street Journal*, 22 October 1993; Robert C. Balling, "Global Warming: Messy Models, Decent Data, and Pointless Policy," in, *The True State of the Planet* ed. Ronald Bailey (New York: The Free Press, 1995), 84–107. See also Robert C. Balling, Jr., *The Heated Debate: Greenhouse Predictions Versus Climate Reality* (San Francisco: Pacific Research Institute for Public Policy, 1992); George C. Marshall Institute, *Scientific Perspectives on the Greenhouse Problem* (Washington, D.C.: George C. Marshall Institute, 1989); and S. Fred Singer, *Global Climate Change: Human and Natural Influences* (New York: Paragon House, 1989).
61. National Academy of Sciences, National Academy of Engineering, and Institute of Medicine, *Policy Implications of Greenhouse Warming* (Washington, D.C.: National Academy Press, 1991); National Research Council, *Global Change and Our Common Future* (Washington, D.C.: National Academy Press, 1989); National Research Council, *Carbon Dioxide and Climate: A Scientific Assessment* (Washington, D.C.: National Academy Press, 1979); Intergovernmental Panel on Climate Change, *Climate Change: The IPCC Scientific Assessment* (Cambridge: Cambridge University Press, 1990); IPCC, *Climate Change: The IPCC Response Strategies* (Washington, D.C.: Island Press, 1991); IPCC, *Climate Change: The Supplementary Report to The IPCC Scientific Assessment* (Cambridge: Cambridge University Press, 1992).
62. Flavin and Tunali, "Getting Warmer," 16.
63. Christopher Flavin, "Climate Policy: Showdown in Berlin," *WorldWatch* 8 (July–August 1995): 8–9; "The Great Berlin Greenhouse Compromise," *Nature* 374 (27 April 1995): 749–50.
64. Intergovernmental Panel on Climate Change, *Climate Change 1995: The Science of Climate Change* (New York: Cambridge University Press, 1996), 4–5.
65. Associated Press, "Nations Urged To Pass Laws On Emissions," *New York Times*, 19 July 1996, A5; Bhushan Bahree, "U.S. Proposes Binding Limits For Emissions," *Wall Street Journal*, 18 July 1996, A5.
66. "The Great Berlin Greenhouse Compromise," *Nature* 374 (27 April 1995): 749–750.
67. Ibid.
68. Christopher Flavin, "Slowing Global Warming: A Worldwide Strategy," Worldwatch Paper 91 (Washington, D.C.: Worldwatch Institute, 1989), 26.
69. Susan Okie, "Developing World's Role in Global Warming Grows," *Washington Post*, 15 May 1990.
70. See World Resources Institute, *World Resources 1994–95* (Washington, D.C.: WRI, 1994), 201.

71. Elena Wilken, "Tropical Diseases in a Changing Climate," *World-Watch* 8 (January/February 1995): 8–9.
72. Anjali Acharya, "Climate Change Endangers the Northern Forests," *World-Watch* 8 (January/February 1995): 8.
73. Hilary French, *After the Earth Summit: The Future of Environmental Governance*, Worldwatch Paper 107 (Washington, D.C.: Worldwatch Institute; March 1992), 10–11.
74. Norman Myers, *Gaia: An Atlas of Planet Management* (New York: Anchor Books, 1992), 76–77.
75. Myers, *Gaia*, 78–81.
76. Ibid., 90.
77. William J. Broad, "Plan to Carve Up Ocean Floor Riches Nears Fruition," *New York Times*, 29 March 1994, B5.
78. Myers, *Gaia*, 91.
79. Ludwik A. Teclaff and Eileen Teclaff, "Transfers of Pollution and the Marine Environment Conventions," *Natural Resources Journal* 31 (Winter 1991): 187–211.
80. Robert D. Hayton, "Reflections on the Estuarine Zone," *Natural Resources Journal* 31 (Winter 1991): 123–38.
81. See Jonathan S. Adams and Thomas O. McShane, *The Myth of Wild Africa: Conservation without Illusion* (New York: Norton, 1992); David Attenborough, *Life on Earth* (Boston: Little, Brown, 1979); David Attenborough, *The Living Planet* (Boston: Little, Brown and Company, 1984); James M. Broadus and Raphael V. Vartanov, *The Oceans and Environmental Security: Shared U.S. and Russian Perspectives* (Washington, D.C.: Island Press, 1994); Paul and Anne Erlich, *Extinction: The Causes and Consequences of the Disappearance of Species* (New York: Ballantine Books, 1981); Robert Goodland, ed., *Race to Save the Tropics: Ecology and Economics for a Sustainable Future* (Washington, D.C.: Island Press, 1990); Edward Goldsmith and Nicholas Hildyard, *The Earth Report: The Essential Guide to Global Ecological Issues* (Los Angeles: Price Stern Sloan, 1988); Suzanne Head and Robert Heinzman, eds., *Lessons of the Rainforest* (San Francisco: Sierra Club Books, 1990); Philip Hurst, *Rainforest Politics: Ecological Destruction in South-East Asia* (London: Zed Books, 1990); Jeffrey A. McNeely et al., *Conserving the World's Biological Diversity* (Gland, Switzerland: International Union for Conservation of Nature and Natural Resources, World Resources Institute, Conservation International, World Wildlife Fund–U.S. and the World Bank, 1990); Norman Myers, *The Primary Source: Tropical Forests and Our Future* (New York: Norton, 1990); Walter V. Reid and Kenton R. Miller, *Keeping Options Alive: The Scientific Basis for Conserving Biodiversity* (Washington, D.C.: World Resources

References

Institute, 1989); Richard Tobin, *The Expendable Future: U.S. Politics and the Protection of Biological Diversity* (Durham, N.C.: Duke University Press, 1990); E. O. Wilson, *The Diversity of Life* (Cambridge, Mass.: Harvard University Press, 1992); E. O. Wilson, ed., *Biodiversity* (Washington, D.C.: National Academy Press, 1988); World Conservation Monitoring Centre, *Global Biodiversity: Status of the Earth's Living Resources* (London: Chapman & Hall, 1992); World Resources Institute, The World Conservation Union, and United Nations Environment Programme, *Global Biodiversity Strategy: Guidelines for Action to Save, Study, and Use Earth's Biotic Wealth Sustainably and Equitably* (Washington, D.C.: World Resources Institute, 1992).

82. Sandra Postel, "Denial in the Decisive Decade," in *State of the World 1992*, ed. Lester Brown et al. (New York: Norton, 1992), 3.
83. E. O. Wilson, quoted in Myers, *Gaia*, 159.
84. Myers, *Gaia*, 26, 41.
85. William Ophuls and A. Stephan Boyan, Jr., *Ecology and the Politics of Scarcity Revisited* (New York: W. H. Freeman, 1992), 129–33.
86. Anjali Acharya, "Plundering the Boreal Forests," *World-Watch* (May/June 1995): 21–29, at 21–26.
87. Clive Ponting, *A Green History of the World: The Environment and the Collapse of Great Civilizations* (New York: Penguin, 1991), 15–16.
88. Myers, *Gaia*, 22–23, 36–38.
89. Elena Wilken, "Assault of the Earth," *World-Watch* 8 (March/April 1995): 20–27.
90. United Nations, *Convention on Biological Diversity*, 1992, Article I.
91. Ibid., Article 16.
92. Ibid., Articles 17,18.
93. Ibid., Article 19.
94. Ibid., Article 21.
95. International Institute for Sustainable Development, "A Summary Report of the Intergovernmental Committee on the Convention on Biological Diversity," 9, no. 6 (22 October 1993).
96. Thomas F. Homer-Dixon, "Physical Dimensions of Global Change," in *Global Accord: Environmental Challenges and International Responses*, ed. Nazli Choucri, (Cambridge, Mass.: MIT Press, 1993), 47–49.
97. World Resources Institute, *The 1995 Information Please Environmental Almanac* (Washington, D.C.: WRI, 1994), 302–03.
98. United Nations, *Geography of Population*, reported in William K. Stevens, "Feeding a Booming Population Without Destroying the Planet," *New York Times*, 5 April 1994.
99. Tim Carrington, "U.S. Efforts Toward Population Control Embroil It in Dispute over Programs," *Wall Street Journal*, 7 March 1994.

100. Ophuls and Boyan, *Ecology and the Politics of Scarcity Revisited*, 44–47.
101. Digby J. McLaren, "Population and the Utopian Myth," *Ecodecision* (June 1993): 60.
102. Ibid., 63.
103. See Paul R. Erlich and Anne H. Erlich, *The Population Explosion* (New York: Simon and Schuster, 1990); Garrett Hardin, *Living Within Limits: Ecology, Economics, and Population Taboos* (New York: Oxford University Press, 1993); Klaus M. Leisinger and Karin Schmitt, *All Our People: Population with a Human Face* (Washington, D.C.: Island Press, 1994); Laurie Ann Mazur, ed., *Beyond the Numbers: A Reader on Population, Consumption, and the Environment* (Washington, D.C.: Island Press, 1994); George D. Moffett, *Critical Masses: The Global Population Challenge* (New York: Viking Penguin, 1994); and Miguel A. Santos, *Managing Planet Earth: Perspectives on Population, Ecology, and the Law* (New York: Bergin and Garvey, 1990).
104. Wallace Kaufman, *No Turning Back: Dismantling the Fantasies of Environmental Thinking* (New York: Basic Books, 1994).
105. Paul R. Erlich and Anne H. Erlich, *The Population Explosion* (New York: Simon and Schuster, 1990).
106. Erich Fromm, *To Have or To Be?* (New York: Harper & Row, 1976); E. F. Schumacher, *Small Is Beautiful: Economics as if People Mattered* (New York: Harper & Row, 1973).
107. Sandra Postel, "Carrying Capacity: Earth's Bottom Line," in *The State of the World 1994* ed. Lester Brown et al., (New York: Norton, 1994), 17.
108. Dick Kirschten, "Woman's Day," *National Journal* (30 April 1994): 1016–17.
109. Barbara Crossette, "Women's Advocates Flocking to Cairo, Eager for Gains," *New York Times*, 2 September 1994.
110. Barbara Crosette, "Vatican Gives Up Battle To Block Population Plan," *New York Times*, 10 September 1994.
111. Quoted in Kirschten, "Woman's Day," 1019.
112. George Melloan, "What's Wirth Worth? Clinton Should Wonder," *Wall Street Journal*, 22 August 1994.
113. Peter Waldman, "Population Conference Fallout Is Feared," *Wall Street Journal*, 2 September 1994.
114. George Weigel, "Where Marriage Is a Dirty Word," *Wall Street Journal*, 26 August 1994.
115. Peter Waldman, "Population Conference Fallout Is Feared," *Wall Street Journal*, 2 September 1994.
116. David Von Drehle, "Population Summit Has Pope Worried," *Washington Post*, 16 June 1994.

References

117. Alan Cowell, "U.S. Negotiators Push For A Truce Over Population," *New York Times*, 5 September 1994.
118. Alan Cowell, "UN Population Meeting Adopts Program of Action," *New York Times*, 14 September 1994.
119. Gustav Neibuhr, "Forming Earthly Alliances to Defend God's Kingdom," *New York Times*, 28 August 1994.
120. John Lancaster, "Muslims Echo Pope's Rejection of UN Population Document," *Washington Post*, 12 August 1994; John Tagliabue, "Vatican Seeks Islamic Allies In UN Population Dispute," *New York Times*, 17 August 1994; Michael Gorgy, "Saudis Scrap Plans to Go To Cairo Conference," *New York Times*, 30 August 1994; "The Sudan Withdraws from UN Cairo Conference," *New York Times*, 31 August 1994.
121. Waldman, "Population Conference Fallout Is Feared."
122. See John H. Cushman, Jr., "U.S. Asks Bishops to Discuss Family Issues," *New York Times*, 10 August 1994.
123. Alan Cowell, "Vatican Fights Plan to Bolster Role of Women," *New York Times*, 15 June 1994; Cowell, "Vatican Attacks Population Stand Supported by U.S.," *New York Times*, 9 August 1994; Cowell, "Vatican Says Gore Is Misrepresenting Population Talks," *New York Times*, 1 September 1994.
124. Alan Cowell, "Is This Abortion?" *New York Times*, 11 August 1994.
125. J. Navarro-Valls, "The Courage to Speak Bluntly," *Wall Street Journal*, 1 September 1994.
126. Wade Greene, "Overconspicuous Overconsumption," *New York Times*, 28 August 1994.
127. UN International Conference on Population and Development, *Program of Action* (Cairo, 1994).
128. Barbara Crosette, "Population Meeting Opens With Challenge to the Right," *New York Times*, 6 September 1994.
129. Barbara Crossette, "UN Meeting Facing Angry Debate on Population," *New York Times*, 4 September 1994.
130. The United States currently spends about half of its family planning funds on grants to other governments; $60 million is earmarked for purchase of U.S.-made contraceptives; $50 million goes to the UN Population Fund; and the rest goes to private organizations. Alan Cowell, "Conference on Population Has Hidden Issue: Money," *New York Times*, 12 September 1994.
131. Alan Cowell, "Despite Abortion Issue, Population Pact Nears," *New York Times*, 9 September 1994.
132. Barbara Crosette, "A Third-World Effort on Family Planning," *New York Times*, 7 September 1994.

133. Tim Carrigan, "Viewing Population as a Global Crisis, Cairo Conferees Have Missed the Point," *Wall Street Journal*, 12 September 1994.
134. Hilary French, *After the Earth Summit: The Future of Environmental Governance*, 10–11, 13.
135. Sherly WuDunn, "Taiwan to Send Nuclear Wastes To North Korea," *New York Times*, 7 February 1997, A1.
136. Charles E. Di Leva, "Trends in International Environmental Law: A Field With Increasing Influence," *Environmental Law Reporter* (February 1991): 10076–77, 10080–82.
137. For a contemporary criticism of international law, see Charles Krauthammer, "The Curse of Legalism," *The New Republic* (6 November 1989): 44–50.
138. Source: World Resources Institute, *World Resources 1994–95* (Washington, D.C.: WRI, 1994), 224.
139. Ophuls and Boyan, *Ecology and the Politics of Scarcity Revisited*, 255–262.

2

Challenges in Achieving the Goals of Global Environmental Agreements

THE 1992 United Nations Conference on the Environment and Development brought together representatives from more than 180 nations in the largest gathering of political leaders in the history of the world. The conference was the culmination of three years of preparatory meetings. It reflected, as was reported in Chapter 1, a new global consensus concerning the importance of protecting the environment and encouraging economic growth in the less-developed areas of the world. Thirty-five thousand people, including more than 110 heads of state, ambassadors, delegates, environmentalists, industrialists, and reporters, came to Rio de Janeiro discuss how to change the life-styles and economic practices of the other 5½ billion people who inhabit the earth. Many were optimistic that the end of the cold war would permit the nations of the earth to come together to address problems that had largely been set aside because of the competition and tension surrounding the superpowers.

The environmental challenges that the delegates and participants had to overcome illustrate the pervasiveness of the problems they pledged to attack. In going to Brazil, most conference participants used modes of transportation that burn fossil fuels and contribute to the threat of global warming. One of the difficult issues that vexed conference planners was whether to use air conditioning in the meetings where delegates were to discuss the threat of CFCs (widely used as a coolant in air conditioning) to the ozone layer.[1] Once the meetings began, participants were inundated with paper—press releases, draft documents, and reports—but no recycling bins were in sight. The delegates were shielded from the poverty and other urban ills that plague Rio (and most of the other megacities of the world). At one moment in the debate, a delegate from an industrialized nation proclaimed that the Rio Declaration, a brief statement of the fundamental principles underlying the conference, should be made available to hang on the bedroom wall of every child. A delegate from Pakistan rose and replied simply: "In my country, most children don't have bedrooms."[2]

The Earth Summit produced an ambitious agenda—two major treaties for global climate change and biodiversity, a statement of principles for the protection of the world's forests, the creation of a new agency to monitor environmental trends, a declaration of broad principles for integrating environmental protection and development, and a detailed agenda of some twenty-seven sets of policies that countries can pursue to preserve natural resources and promote sustainable development. Although the summit was a remarkable attempt to focus world attention on the health of the earth and the plight of its poorest residents, it was largely a set of promises national leaders made to each other, to the rest of the world, and to generations yet unborn. Such promises have been made before, but never on such a grand scale. Whether these promises will ever be fulfilled remains to be seen.

Devising and implementing national and international environmental policies to respond to the world's natural-resource and pollution problems is an unprecedented political challenge. While scientists, policy makers, the media, and the public continue to debate and explore the nature of the environmental crisis, we need a parallel discussion of how we should act to remedy these problems, once we decide to take action. The prospects for effective policy making in response to global environmental problems are uncertain, particularly given the increasing skepticism of some observers over whether national governments and international bodies have the institutional capacity to take effective collective action.[3]

This chapter focuses first on some of the general challenges in implementing public policies. The discussion then turns to four tasks inherent in achieving the goals of global environmental agreements: integrating environmental and other policy goals, facilitating participation by nations, creating incentives for compliance, and assessing progress and making adjustments. Subsequent chapters address these and related issues in more detail. The assumption underlying all of these issues is that implementation can occur in an incremental fashion, as typically happens, and still be sufficiently effective to accomplish the specific goals of global agreements and preserve the environment for future generations. However, it is not at all clear whether the slow, incremental nature of the policy-making process is sufficient to address the tremendous policy changes that global environmental problems seem to demand. Whether this underlying assumption is too optimistic is the topic of the final chapter.

Implementation and the Policy-Making Process

A host of factors help determine the success of a policy effort—how well the problems to be remedied are understood, the level of public support for change, the commitment of political elites and leaders to take action, the nature of the relationships between polluting industries and governments, the kinds of statutory and administrative approaches to environmental regulation, and many others. In some ways, policy implementation is the culmination of the policy process—the efforts to define problems and formulate effective policy responses are put to the test. But for global agreements, it is also the beginning of a new policy process: National institutions and interests must find ways to implement the provisions of international agreements through formulating and putting in place their own national and subnational laws, programs, and regulations.[4]

Several images have been evoked to help explain the evolution of public policies. One metaphor for policy making is that of policy soup: proposals float in a soup of ideas; they are constantly stirred and changed as they mix with other initiatives; and those that are sufficiently softened and become more acceptable to policy makers eventually emerge from the soup. A "garbage can" model posits that proposals may be offered prematurely and are dumped, only to be retrieved later and recycled when the political environment is more hospitable. The "policy window" model recognizes that windows of opportunity occasionally open, as a result of an event that captures or focuses people's attention and pushes the issue through the window.[5] Models of the policy-making process range from intricate elaborations of the myriad details that comprise policy efforts to rather simple structures of basic elements. In general, the policy process includes several steps: identifying and defining the problem, generating sufficient interest and support to place the issue on the policy agenda, formulating the specific provisions of the policy, enacting the policy provisions and securing the required funds, implementing the policy, evaluating the appropriateness of the policy goals themselves as well as how effectively they are achieved, and then reformulating the policy and beginning the process anew.[6] The process becomes more complicated at the international level. The basic elements of the policy-making process for international environmental problems can be outlined as follows:

1. defining the nature of the problem, and placing it on the international policy agenda;

2. formulating a policy and negotiating an international agreement to be signed by participating states;

3. ratifying the agreement by the parties through their own national political requirements for ratification;

4. formulating, enacting, and implementing laws and regulatory programs by parties to ensure they meet their commitments;

5. monitoring implementation efforts by national and international authorities, reporting to the body responsible for supervising compliance; and

6. evaluating progress, negotiating protocols to the agreement if required, and revising domestic implementing laws and programs.

Most policies are incremental rather than comprehensive, primarily a series of marginal adjustments to earlier efforts rather than dramatic departures from past practices. Although many scholars have defended such an approach as reasonable, given the limitations of policy analysis and the difficulties in mobilizing political support for more fundamental changes, it may produce policies that ultimately do not resolve the problems at which they are aimed.[7]

Policy Definition

Identifying and defining the problem is the first step in the policy-making process. As the nature of the problem is perceived and defined, interested groups and individuals organize in anticipation of presenting demands to government officials. How problems are identified and defined is critical. The perception of the problem might be flawed; attention might be directed toward symptoms rather than root causes. The proposed policy response might not reflect an accurate understanding of the problem. The level of political support generated during this initial stage plays an extremely important role in policy development, implementation, and evaluation. Problems that fail to attract sufficient support might not be addressed, while other, less serious problems receive attention. Perhaps most important, how the problem is defined, who provides the definition, and what assumptions and values underlie the chose definition will have enormous implications for the course the policy takes. United Nations–affiliated bodies, for example, have played the dominant role in shaping the global environmental agenda by deciding what issues will be addressed in international meetings and negotiations. But the agenda has also been shaped by the efforts of nongovernmental organizations and the results of media attention.

Policy Formulation and Enactment

The development of a policy to respond to the demand for action involves formulating a program, getting on the policy agenda of the appropriate governing body, enacting legislation to authorize the program or plan, appropriating sufficient funds, and providing for a mechanism to implement the provisions. These actions are interdependent and not easily separated. The coalition-building and compromising that are central to fashioning agreements often result in vague, imprecise commitments. Difficult choices might be deferred to those implementing the policy, thus postponing the political controversy and conflict from the formulation stage to implementation. Again, United Nations agencies have played a critical role in providing a forum for bringing parties together to negotiate global environmental agreements.

Ratification

Ratifying treaties is primarily a function of domestic politics, institutions, and processes; but the national debate is often colored by concerns about what is happening abroad, how ratification or nonratification will affect international relations, and how serious the problem is believed to be. In the United States, as in other countries, ratification of global agreements is usually intertwined with domestic politics, and agreements have sometimes been blocked as a result of concerns that are quite unrelated to the proposed treaty. International pressure may have some impact on ratification, but just as often, political leaders may find some benefit in taking a tough line against foreign pressures in order to pursue what they argue to be their nation's self-interest. Indirectly, international pressure can be brought to bear through domestic organized interests and expressions of public opinion, so that ratification is demanded by citizens who insist that their nation contribute to a global effort. Ratification of global environmental agreements is largely a function both of governmental efforts to foster global cooperation and of calls for action by environmental interests and the media.

Implementation and the Policy-Making Process

Implementation

Implementation can be defined simply as the carrying out of public policies and the ability to achieve desired outcomes.[8] It is often a long, complicated process of interpreting the intent of those who drafted the convention or treaty, creating administrative structures and processes, reopening the policy formulation debates as laws and regulations are devised, and building political support for enforcing the new obligations. It involves both domestic and global administrative bodies. Domestic governing institutions must formulate laws and create administrative bodies to implement them and to monitor compliance; international agencies may be responsible for collecting and disseminating data in order to encourage compliance with treaty mandates, verify compliance, provide technical and financial assistance, resolve disputes among parties, and impose sanctions.

Monitoring Compliance

The success of global efforts depends on the availability of accurate and timely information on the extent to which parties are implementing and complying with treaty provisions. Information gathered by individuals and private organizations can be provided to the agency responsible for monitoring compliance, as well as to governments, nongovernmental organizations, the media, and others. These data are of major importance in generating pressure on states to keep their commitments and in assessing the effectiveness of treaties. Studies of international agreements regularly emphasize the importance of collecting timely and accurate information and conclude that the quality and quantity of this information generally must be improved to know how well policies are being implemented.

Evaluation

The last step in the policy process is evaluation, but it is somewhat misleading to characterize this as a separate step, since it permeates the entire effort. While it is expected that the evaluation will be

CHALLENGES IN ACHIEVING THE GOALS

politically neutral, analytic, and professional, policy evaluation is very much a political undertaking, pursued by policy makers for a variety of purposes. There is constant tension between careful policy analysis and political calculations. Policy efforts are constantly being evaluated and alterations in programs proposed. Policy evaluation is also difficult to do well. Policy makers are often flooded with information, but little of it may be in useable form. Policy analysis cannot resolve conflicting values at the heart of many policies. Conditions constantly change with new information and experience, and adjustments to original analyses are essential.

This model of the policy-making process is limited as an analytic tool. It does not explain or predict why policies take the shape they do. The actual process does not neatly fit these six categories, and the steps often overlap. There may be little agreement as to the definition of the problem; some policies may be placed on the policy agenda by accident or by a failure to act; implementation usually triggers an identification of yet another set of problems, and efforts to solve one problem usually produce unanticipated consequences that trigger a new round of policy-making efforts.[9] Focusing on the discrete elements of each step may disguise broader developments. Nevertheless, the model provides a useful way to structure an examination of the importance of implementation and how it interacts with the other steps in the policy-making process.

Challenges in Implementing Environmental and Other Public Policies

Implementation is a critical step in the policy-making process. If agreements are not effectively implemented, then the entire effort to achieve cooperation or bring about change is threatened. The prospects for effective implementation can be assessed a number of ways. A straightforward inquiry asks the extent to which the formal goals, set forth in statutes or treaties, are achieved. Success in achieving public policy goals lies in the details: among the most critical concerns are whether sufficient authority and resources are provided to accomplish the mandates, and whether effective incentives are created to encourage compliance. Another form of assessment looks at implementation in terms of how well it satisfies core political values such as social justice or equality.[10] From yet another perspective, fairness in the implementation process or representation of all interests may be central to the assessment. The study of implementation directs attention to how well governments can accomplish their goals, how they can improve their capacity to implement policies, and how policy formulation can better plan for effective implementation.

In reality, the implementation of public policies has pitfalls in every stage. The policies themselves, as expressed in statutes, are often vague and unclear. Allocated resources may not be adequate; the expectations to which statutes give rise may overwhelm the political will to accomplish the goals identified. In the United States, separation of powers has made implementation difficult because the legislature may be of a different party than the executive-branch officials who are responsible for implementation. Implementation is further complicated by federalism: Most national policies are implemented by state officials who have their own political agenda and policy priorities or who may be interested in ensuring their state's autonomy and independence. Policy statements such as statutes often give little attention to the realities of implementation by administrative agencies or to the importance of creating effective incentives for compliance by regulated parties.[11]

These problems are greatly magnified when international agreements are to be implemented. Treaties or international agreements may suffer even more from ambiguity and impreciseness because the political process surrounding global agreements is so much more complicated than for domestic issues. Nation-states will be particularly sensitive to perceived threats to their sovereignty arising from international regulatory regimes. The existing regulatory frameworks responsible for carrying out global agreements will vary widely and reflect differing commitments to environmental protection. Issues of fairness, ability to pay, impact on competitiveness in international markets, national security, and a host of other difficulties greatly complicate efforts to implement international agreements. Several characteristics of implementation are particularly important to an understanding of the prospects for effectively accomplishing the goals of global environmental accords, and are discussed below.

Implementing a Moving Target

Implementation is a dynamic, ongoing enterprise. "Implementation is evolution," write Pressman and Wildavsky; "it will inevitably reformulate as well as carry out policy."[12] Implementation is not limited to agencies and bureaus, but is broadly defined to include

> the web of direct and indirect political, economic, and social forces that bear on the behavior of all those involved, and ultimately the impacts—both intended and unintended—of the program.[13]

Such problems as improving environmental quality and protecting natural resources, are rarely "solved": implementation is an effort to try and manage them, make some incremental progress in reducing their consequences, and adjust policy efforts in light of experience. Policy choices are often revisited and revised during implementation, and adjustments in institutions and strategies must be made at the expense of developing a consistent and coherent program. Implementation efforts are often at the mercy of changes in the nature of the problems or the political support to respond to them. New political leaders may alter goals. Initial energy and commitment may lag. As implementation efforts proceed, we become aware of new problems, constraints, and opportunities. Resources and goals are constantly in flux; unintended consequences are inevitable.[14] Sometimes implementation fails from a lack of political will or agreement or from inadequate funding or authority. Environmental goals may conflict with each other, and with other policy objectives, for limited resources. Political compromise often results in ambiguity, and there may be little agreement over what exactly the

Challenges in Implementing Public Policies

goals are. Policy goals may fall victim to inadequate implementation or unrealistic objectives. Goals and the means provided to accomplish them could be seriously mismatched. The implementation of public policies is highly dependent on what occurs elsewhere in the political process. In the United States, implementation is usually a complicated process of interpreting congressional intent, merging statutory and presidential priorities, creating administrative structures and processes, revisiting the political debates of the policy formulation stage as administrative regulations are devised, and building political support for implementing and enforcing regulatory requirements. This model of the policy process assumes a simple relationship between policy formulation and implementation, but the lines between making and implementing policies are heavily blurred, and those who implement laws are constantly confronted with policy choices. Implementation of international agreements may begin before the formal ratification process begins, to ensure that enforceable provisions required to bring a country into compliance with expected obligations will be in place when those obligations are formalized.

Incremental versus Comprehensive Action

At one level, implementation involves assessing the effectiveness of alternative policy tools, mechanisms, and approaches. Policy analysts compare the advantages and disadvantages of proposed regulatory schemes over subsidies and traditional, centralized, command-and-control regulatory programs over decentralized, market-based approaches; they examine alternatives in the level of resources required to bring about certain policy responses; and they explore the interaction of different policy efforts. Much of the challenge of implementation lies in matching the agreed-on policy goals with the appropriate policy tools. Global environmental problems can be addressed at this incremental level. Economic incentives can be created to reduce the loss of biodiversity, for example. Innovative trading programs can be devised to reduce the cost of greenhouse gas reductions. Governments can ban production of ozone-depleting chemicals. From this view, implementation is firmly rooted in the realm of the possible.

If implementation remains on this plane, however we will treat some symptoms and buy some time, but we will not likely solve core, underlying problems. Even more dangerous, we may delude ourselves into thinking that we have taken sufficient action until it is too late and catastrophe is on us. Many people believe environmental problems are fundamentally rooted in modernity, technology, industrialization, economic growth, the pursuit of wealth, population growth and consumption, and the idea of progress. Tinkering with prices, end-of-pipe pollution-control equipment, trading pollution allowances, and other such incremental policy tools are doomed to fail in the long run. Fundamental transformations are needed in industrial activity, energy production, transportation, and in attitudes about consumption, convenience, quality of life, and individual choice.

One of the great challenges of implementation is in fashioning creative ways for these two levels of change to interact. Incremental changes are critical because they are the most politically feasible. But these modest steps must be directed toward the major shifts in attitude and behavior that may be required, eventually, to respond effectively to the serious environmental problems we face. We have little experience in initiating and shaping genuinely fundamental social changes. Perhaps they are not possible unless catastrophes or crises occur. Short of that, we simply may not be able to make major changes in our economic lives, no matter how sobering the scientific data or clear the consensus. Nevertheless, we need to continue to debate large-scale changes and explore their possibilities, while at the same time moving quickly to implement feasible changes that will move us in the right direction and provide a foundation for future actions.

Although ambitious, comprehensive policy goals may appear to be unavoidable response to global threats, the failure to achieve them breeds cynicism, discouragement, and contempt for collective efforts and for the power of government to remedy public problems. The health of the body politic is enhanced by fewer, more modest goals that we are more likely to achieve, rather than un-

Challenges in Implementing Public Policies

realistic aspirations to which we are not really committed. Major discrepancies between public goals and policy implementation can pose a serious threat to the legitimacy of a political regime.[15] Implementation involves learning to identify what changes are necessary and then make steady progress in accomplishing them, and at the same time avoiding the temptation to proclaim broad policy goals we do not intend to pursue. Developing the capacity of governments to function efficiently and effectively is just as important as making direct progress in improving environmental quality, protecting natural resources, reducing poverty, and meeting other global challenges. Part of implementation is assessing the impact of policy efforts on the policy process itself and on our capacity, over time, to accomplish important public purposes.

Limited Governmental Capacity

Our limitations in remedying such major social problems as environmental degradation are rooted in the policy-making process. In the United States, major obstacles to implementation are the separation of powers and government gridlock. From a constitutional perspective, policy making in the American political system was in many ways designed to be inefficient, to check the power of policy makers, and to permit action only with widespread support among different constituencies and their representative institutions. What was once defended as a virtue, however, is now regularly derided as a vice. Deliberative democracy has been replaced by deadlock. Many people believe that institutional barriers to coherent and effective policy making prevent the federal government from responding to the challenges confronting the nation. These people fear that divided government—with Congress and the White House controlled by leaders of different political parties whose primary goal is the political defeat of the opposition—makes good, effective policy formulation extremely difficult at best. Other institutional and structural characteristics of American government, such as frequent elections, impede development of coherent, long-term strategies to resolve and prevent problems.[16]

Such critics as Theodore Lowi have argued that implementation's chief shortcoming is that it allows well-organized interests ultimately to corrupt and weaken the policy-making process. Policy-making discretion in implementing agencies threatens democratic formalities of accountability, responsibility, and, ultimately, the rule of law and individual freedom. Cynicism and criticism are the outcome when public power is subverted for private ends, and governments at all level are seen as incapable of planning or taking effective action to resolve problems.[17]

Charles Lindblom and Edward Woodhouse argue that many of the problems we are concerned about are so complex that we cannot understand their causes or come up with solutions.[18] We lack effective political mechanisms to engage the public in a real debate over alternative perspectives and views. Although we have made some progress in measuring social problems and identifying their causes, it has been much more difficult to compare alternatives and develop effective policy interventions. Analysis can help identify choices, but most policy questions are ultimately political ones involving fundamental conflicts among competing social values.[19]

Other skeptics contend that there is a tradeoff between democracy and reasoned analysis, between government by representatives and by experts, and it is not clear how we can have more reasoned democratic decision making.[20] Solutions may appear simple, but coordinating policies is often daunting, and the complexity of joint action makes most policy efforts seem impossible. Although the probability of individual steps being taken may be rather high, the cumulative odds of every required action occurring become astronomical.[21]

The Tractability of the Problem

Mazmanian and Sabatier argue that the characteristics of the problems to be remedied are critical in determining the success of implementation.[22] Some environmental and natural resource problems are, of course, much more difficult to address than others. They may require new, cleaner technologies that may or may not be available, or may be too expensive for many users. New technologies may require the regulation of a wide range of diverse and geographically dispersed behavior as well as major modifications in local practices. Mazmanian and Sabatier argue that problems are most likely to be remedied if the following conditions are present:

1. A valid theory connects behavioral change to problem solution, the requisite technology exists, and measurement of impact is inexpensive.

2. In the behavior that causes the problem has minimal variation across its geographical range

3. The target group is an easily identifiable minority of the population within a political jurisdiction.

4. The amount of behavioral change required is modest.[23]

However, one reason we study implementation is to improve our capacity to remedy relatively intractable problems, so even if all these conditions are not present, implementation is not necessarily impossible. It is just more difficult, and advocates will need to generate additional political support.

The success of implementation depends on a number of other factors, including the strength of the perception that the policy is needed, the nature of the interests threatened, public opinion, and historical accidents and antecedents. Perception of the seriousness of the problem may vary over time as other public concerns clamor for attention. Differences in local conditions require flexibility and adaptability. The more powerful an industry is, the greater its ability to resist regulation. Changes in the commitments, concerns, and resources of public interest and other groups also determine how much government officials are pressured to implement laws and regulations. The level of scientific agreement about the causes of the problems is another key factor: The clearer and more compelling the scientific research, the greater the likelihood of political support for change. If a clear causal theory connects policy efforts with the resolution of the problem, implementation will more likely be successful.

Political and Institutional Support for Implementation

Some policy statements will have greater political support than others. A strong statutory endorsement of an environmental policy, for example, has greater legitimacy than an administrative regulation or a judicial decision. Clear, precise, unambiguous policy authorizations can increase the likelihood of implementation. The more the new policy requires a departure from existing commitments and practices, the more difficult implementation will be. The provision of adequate resources is an obvious requisite for policy implementation.

The integration of policy efforts across the agencies involved is critical; the lead implementation agency must be able to work with other national agencies as well as subnational ones. Certain policy decisions may make it easier or harder to accomplish the intended goals. Procedural requirements, for example, can slow down implementation, divert resources, or provide opportunities for opponents to defeat initiatives. Change in behavior is unlikely unless agency officials are clearly committed to achieving the policy goals. New policy interventions may require special leadership efforts to develop new regulations and procedures and to enforce them. Statutes can also shape the influence that outside groups have on accomplishing policy objectives. Judicial standing, for instance, may be given to parties to challenge administrative actions or compel performance of nondiscretionary duties.[24]

The "creation of organizational machinery for executing a program" is also a crucial step.[25] Organizations are not neutral vehicles for policy implementation—rather, their structures and processes affect efforts, and their officials have their own personal and policy agendas. Some organizations are better than others at providing the flexibility and resiliency need for certain kinds of policy implementation. Some organizations may be good at adapting, while others are too rigid to make necessary adjustments. Implementation can be regarded as either a learning effort, or a mechanized, routinized way to accomplish a goal: the more complicated the task, the less likely the latter kind of organization will be effective.

Mazmanian and Sabatier suggest that for implementation to be effective, policies should include:

1. clear and consistent objectives;
2. substantive criteria for resolving goal conflicts;
3. a sound theory that identifies the principal factors and causal linkages affecting policy objectives;
4. sufficient authority given to implementing officials over target groups and other points of leverage;
5. assignment to sympathetic agencies with effective structures, sufficient financial resources, and strong political support;
6. delegation to administrative bodies with managerial and political skills, commitment to policy goals, effective structures, and sufficient financial resources;
7. strong support from organized constituency groups and key policy makers (legislators or chief executives); and
8. objectives with sufficient priority that they will not be undermined over time by the emergence of conflicting public policies.[26]

These conditions were identified in terms of implementing domestic policies, but they also provide a useful checklist for global agreements. All these conditions are not likely to occur for any one policy, but they illustrate how challenging it is to devise effective means of implementing public policies.

Challenges in Implementing Environmental Agreements

Few studies of the implementation of global environmental agreements have been conducted, as the focus of attention has naturally been aimed at the challenges that arise in the formation of these accords. But some studies of existing agreements, as well as broader inquiries into implementation, have begun to emerge. The International Institute for Applied Systems Analysis began a three-year study in 1994 of the implementation of international environmental agreements.[27] The study focuses on the interaction of international agreements and domestic politics in both directions—how domestic concerns affect implementation of international agreements and how the realities of implementation affect the content of international agreements. In identifying key variables that contribute to effective implementation, the study centers on two questions:

1. How does participation in and access to the process of implementation at the domestic level affect performance of international commitments?

2. Under what conditions does the content of an international agreement co-evolve most closely with the reality of what states can implement domestically?

The primary challenge of implementation is how to encourage effective compliance by countries that are parties to the agreements. That broad issue, in turn, raises a number of additional concerns. Given the inevitable domestic economic and political barriers, for example, the international community must create incentives to ensure that participating nations implement programs that achieve the goals set forth in global agreements. Without any means of encouraging and pressuring nations to implement the agreements, implementation will constantly be threatened, as when some countries calculate that they can avoid compliance costs and perhaps gain competitive advantage for domestic industries. Other countries will comply with agreements as a means of demonstrating leadership or through the far-sightedness of their leaders. To bolster such motivations, agreements will need several kinds of provisions. They must provide for a global reporting system so that emission levels can be compared with limits included in the agreements. An international body will be required to monitor reporting data and to impose sanctions when compliance agreements are not satisfied. Economic sanctions must be imposed when implementation failures occur. They may range from cutoffs of foreign assistance to import barriers on goods produced in noncomplying countries.

Challenges in Implementing Environmental Agreements

Some international body will have to be empowered to decide disputes. As indicated above, financial and technological assistance must be provided to the poorer countries to encourage compliance or simply to make it possible. These sanctions and incentives are of limited power, since they require unparalleled cooperation and unity on the part of the nations of the world. The process of formulating global agreements itself will be critical in building support for the measures agreed to by ensuring that all parties feel their interests were considered and share a commitment to making the agreement work.

The UNCED secretariat analyzed some one hundred international environmental instruments and offered thirty-two criteria in six broad categories for assessing the effectiveness of implementing agreements.[28] That and other studies of implementation that are discussed below suggest the following set of questions for examining what is involved in implementing global environmental agreements:

1. Integrating Environmental and Other Policy Objectives
   a. How well are different environmental goals integrated and coordinated?
   b. Do goals integrate environment and development and otherwise take into account the special circumstances of less-developed countries?
   c. How are environmental goals integrated with other important policy concerns?
2. Encouraging Participation
   a. What incentives are available to encourage participation and facilitate implementation?
   b. What incentives are particularly aimed at the less-developed countries, such as the transfer of new technologies that are clean, efficient, and effective in remedying pollution, and financial assistance?
   c. To what extent is participation in policy debates and formulation plans by nongovernmental organizations encouraged as a way to broaden support and encourage domestic pressures to comply with agreements?

CHALLENGES IN ACHIEVING THE GOALS

3. Creating Incentives for Compliance
   a. What commitments are imposed on parties? How is compliance with these commitments monitored and measured?
   b. How do parties report on their performance in implementing agreements? What are the requirements for providing data?
   c. What resources, other incentives, and disincentives are provided to encourage compliance with mandates?
   d. How is information on implementation made available to governments, industries, and the public?
4. Promoting Compliance, Evaluation, and Adjustment
   a. What mechanisms are available to remedy disputes over implementation and how do they accommodate participation by national authorities, subnational governments, and NGOs?
   b. To what extent are periodic reviews and assessments of implementation ensured, so that agreements can be revised and updated when necessary?
   c. How is new knowledge incorporated into policy formulation and implementation?
   d. Which existing institutions are well suited to performing essential functions such as monitoring compliance and what additional institutions might be needed?

There are no precise boundaries between these questions. Issues cut across these rough categories. But they provide a framework for exploring the issues involved in implementation and are explored in some detail below and in the next four chapters.

Integrating Environmental and Other Objectives

Despite their diversity, global environmental problems share some important difficulties in formulating and implementing remedial policies. The problems are intricately interrelated, and implementation efforts must be carefully coordinated. Ozone and other urban

Challenges in Implementing Environmental Agreements

atmospheric pollutants, for example, pose immediate health risks to humans in addition to acting as greenhouse gases. Chlorofluorocarbons damage the stratospheric ozone layer and contribute to the threat of global warming. The burning of fossil fuels threatens human health and is also a precursor of acid rain.[29] Efforts to reduce one form of emissions (acid rain emissions through the use of scrubbers, for instance) may increase other kinds of emissions (greenhouse gas emissions may grow as more coal is required to produce the same amount of electricity), rather than reducing total levels.

Given this interaction among ills, solutions to environmental problems do not simply require new technologies but a much broader, integrated response. Urban air pollution, for example, cannot be remedied only by a new round of tailpipe standards. Effective solutions require changes in traffic patterns, land use planning and development, social and cultural practices, economic incentives, and public investment in mass transit and other transportation infrastructures. Similarly, reducing the likelihood of global climate change will require not only new technologies, but will also require widespread changes in behavior.[30]

Progress has been made in reducing or eliminating some forms of pollution, but major challenges remain. This is particularly true in the industrialized nations, where environmental laws and programs have been in place for a number of years, and where levels of traditional pollutants have in many cases been reduced. However, those pollutants that remain are the most difficult to mitigate, and the costs of reducing the final increments are much greater than those required to achieve earlier reductions. The development of new chemical and nuclear industries and technologies poses a further problem in that relatively minute levels of emissions can result in significant environmental and health hazards. Some people argue that pollution will soon be a thing of the past in the industrialized world, that economic growth and environmental quality are mutually reinforcing goals, and that a growing population poses no threat to the global ecosystem.[31] However, such optimism seems unwarranted given our limited understanding of the global environment.

CHALLENGES IN ACHIEVING THE GOALS

Implementation of global environmental policies cannot be made without regard for other social goals. Environmental concerns, important as they are, must compete with other demands for action. Among the wealthy nations, the relentless demand for economic growth and increased consumption places tremendous pressure on the earth's biosphere. In the United States, middle-class fears about an uncertain economic future are among the most potent political concerns. Among the poorer nations, pressures to preserve the environment cannot cancel out demands for improving the quality of life for the $1^{1}/_{2}$ billion people whose lives are mired in poverty. Policy efforts aimed at protecting environmental quality and natural resources must be coordinated with other policies such as economic development in the Third World, international trade, human rights, and national security (see Chapters 5 and 6).

The challenges confronting the global community in implementing international environmental agreements are many. To identify and discuss them is to propose an agenda for actions to remedy them. One of the elements of the debate is to recognize the interrelatedness of problems and solutions. A list of the factors that affect the environment, for example, would likely include energy consumption, technological development, food needs, population growth, population movement, national and international trade, economic production, national debt and wealth, changing land use, consumption of natural resources, urbanization, and industrial and household waste.

Crucial to implementing environmental accords is balancing the need for a comprehensive, integrated approach with the threat of being overwhelmed by the complexity. The more complicated the model of the causes of and remedies to global predicaments, the more bewildering may be the decision where to begin or where to intervene. Implementation ultimately requires linking environmental variables and processes to economic, political, and social actions; devising appropriate policies to deal with the global environment; and creating effective institutions to ensure coordinated global responses.

Encouraging Participation

THE "TRAGEDY OF THE COMMONS"

Overexploitation of common resources is an inevitable outcome of the exercise of rational self-interest, given the incentives to use these resources before others do. To prevent overexploitation, rights and duties must be devised to protect these resources. Virtually all ecological resources—local airsheds, the global atmosphere, watersheds, oceans, and the biosphere itself—are resources held or used in common. Pollution is also a commons problem where the incentives lead to pollution, since the share of environmental damage suffered by the polluter is less than the cost of preventing it. The metaphor of the tragedy of the commons "is not merely an assertion of humanity's ultimate dependence on the ecological life-support systems of the planet; it is also an accurate description of the current human predicament."[32] The problem of the commons operates internationally as well as locally, as pollution of the atmosphere and oceans increases and global resources decrease. The international challenge to safeguard the commons becomes more profound, because nation-states guard their sovereignty jealously and internationally binding agreements are so difficult to enact and enforce. We cannot escape self-interested motivations; but we may come to realize that it is in our self-interest to protect local and global ecosystems by making the changes that will be required.

Some argue that protecting the atmosphere or the oceans requires the creation of a public good that incorporates incentives for all nations to contribute, rather than permitting some to be free riders—enjoying the good without contributing to its preservation. Elinor Ostrom and others have studied how some communities are able to devise decentralized means of establishing and enforcing limits on the use of common resources.[33] It is far more difficult to manage the atmosphere and oceans, where a decentralized approach is not possible.

Some international relations scholars have divided international environmental problems into two types, collaboration and coordination, that highlight additional challenges of implementation. Collaboration problems are more difficult because they involve cases where vital but conflicting national interests are at stake. Pollution reduction is a collaboration challenge, for example, since controlling pollution may impose costs on business that impair their international competitive position or threaten sovereignty. Coordination problems are easier to solve because all states benefit from collective action, but they may be reluctant to collaborate because of inertia or other factors, or may not be able to agree on what course of action to follow. Table 2.1 classifies major treaties along this dimension.

TABLE 2.1
KINDS OF GLOBAL ENVIRONMENTAL ACCORDS

Kind of problem	Coordination		Collaboration		Total Accords
Air pollution	0	0%	6	100%	6
Conservation	17	41%	24	59%	41
Framework	3	20%	12	80%	15
Land-based water pollution	1	13%	7	88%	8
Marine dumping	1	14%	6	86%	7
Nuclear regulation	7	88%	1	13%	8
Oil pollution	5	28%	13	72%	18
Other	8	67%	4	33%	12
Plant disease	9	100%	0	0%	9
Worker protection	8	100%	0	0%	8
Total	59	45%	73	55%	132

SOURCE: Peter M. Haas with Jan Sundgren, "Evolving International Environmental Law: Changing Practices of National Sovereignity," in *Global Accord: Environmental Challenges and International Responses* (Cambridge, Mass.: MIT Press, 1993), 401–29, 412.

International conventions are needed to help states collaborate in their collective self-interest. Countries can hold each other hostage, if one country fails to act, others bear some of the consequences; solutions require the participation of all the major sources of the

Challenges in Implementing Environmental Agreements

problems.[34] Some kind of international policy-making body with real authority to establish and enforce standards also seems to be requisite for protecting the global commons. But local, grassroots efforts that preserve resources while meeting material needs are also essential. Can environmental problems be addressed by current institutions and structures at all levels of government, so that the challenge is primarily one of coordination? Or do we need to make qualitative changes, create new institutions and processes? Are environmental challenges so difficult or complex that they require responses that transcend existing international institutions?

INSTITUTIONS TO ENCOURAGE PARTICIPATION

Nothing short of global government, some argue, will resolve current global problems. Others hold that national sovereignty need not be abolished, but that nation-states can pursue global cooperative efforts voluntarily. Some call for strong, centralized agencies to implement global agreements, while others call for more decentralized mechanisms. Efforts to strengthen the institutional capacity of governments to regulate behavior and direct efforts within their own boundaries and to interact with others is essential, but whether further steps toward global governance are necessary or possible remains unclear.

A particularly serious difficulty in implementing global agreements is the constraint posed by national sovereignty. If global accords are to be more than appeals to voluntary efforts, they must provide for the enforcement of obligations. But the commitment to national sovereignty is a powerful one; despite the great attention focused on the state of the global environment at the 1992 Earth Summit in Brazil, for example, the participating nations confirmed the right of each nation to control the activities within its border, even when those activities threaten global well-being. Leading countries are influential in mobilizing others to participate in these treaties. These agreements have incrementally eroded the strength of national sovereignty, but there has been no major transformation in international law, only a gradual expansion of international agreement and cooperation.[35]

Solutions to worldwide problems bump up against other local tendencies. Nongovernment organizations are a growing force, demanding participation along with official national delegations in global meetings and negotiations. Skepticism of centralized, bureaucratic policy making is on the rise, as evidenced by recent elections in the United States and the other industrialized democracies as well as by the fall of communism in the early 1990s. Democracy can create pressure for improved environmental protection, as is happening in Eastern Europe and as has occurred in Western Europe, North America, Japan, Australia and New Zealand, but it can also obstruct the development of institutions to facilitate democratization and effective policy making.

INCENTIVES TO PARTICIPATE

Countries may choose to participate in global agreements for a number of reasons. Participation may be in a nation's self-interest, and agreements themselves are negotiated to further self-interest and pressure other states to change their behavior. Agreements may prohibit actions that no states currently undertake as a way to restrain future actions that might be viewed as harmful. The specific kinds of agreements negotiated may also be an opportunity to pursue self-interest. Treaty provisions that are vague and indeterminate, for example, may permit states to continue current behavior while claiming that they have conformed to international standards. Parties may agree only to provisions that they have already decided to undertake. They may use international forums to generate support and legitimacy for actions that may be unpopular at home.[36]

Countries may take a broader, more long-term view of self-interest and engage in agreements that go beyond short-term, narrow interests. Uncertainty, possibly adverse consequences may compel action that would otherwise appear inconsistent with more immediate and obvious incentives. Countries may see it in their long-term self-interest to be leaders in environmental protection, or may fear the consequences of being seen as following others. Nongovernment actors, including multinational and domestic corporations, scientific groups, environmental interests, and others can help

Challenges in Implementing Environmental Agreements

shape a demand for signing and implementing international agreements. These are often dynamic forces whose relative strength changes through political and electoral cycles. New scientific findings, the discovery of alternative technologies, and other developments affect participation.[37]

Participation is affected by other calculations. Nations may conclude that their participation will be matched by similar actions by other states. Interests may be sufficiently complementary that there is little need for enforcement. States may recognize that collaboration is necessary to produce public goods or protect common resources. However, even where joint participation is preferred, states may view it in their own interest to violate agreements as others continue to comply with them. In such cases, enforcement is essential to avoid free riders and compel compliance.[38]

Participation may be encouraged a number of ways. Adequate funding of compliance efforts is a chronic problem of collective action. Financial incentives can be offered; states can transfer compliance-related funds to poorer nations. Educational and training programs on the need for regulation, alternative means of complying with obligations, and other topics can foster participation and help generate support for policy efforts. Bureaucratic routines may be hard to interrupt once firmly entrenched. Countries do not constantly reassess their interests and power and become habituated to certain actions, even when their self-interest would dictate a reassessment or even a change. However, participation is not necessarily static, but can be affected by a host of factors such as changes in scientific understanding, technological solutions, economic growth or recessions, and elections or other changes in governance.[39]

Countries may be pressured by domestic and international sources to sign an agreement, but may then decide not to participate. The costs of participation may outweigh the benefits received. Countries may want to enjoy the benefits of the agreement as free riders, without incurring the obligations. More pressing concerns may divert the resources intended for participation. Other countries may have a sincere intent to implement agreements, but they may lack the resources or the technical and administrative capacity to do so. Countries may make good faith efforts, but fall short in compelling the necessary cooperation from industries, local governments, and individuals.[40]

Domestic politics and policies also shape the participation of nation-states in international discussions and shape the kinds of initiative they will support and the kinds of policies they will implement. Although some countries may be motivated by international norms, threats of sanctions, or other external considerations, nations generally respond to domestic political pressures. Domestic political influence aimed at protecting entrenched interests is difficult to overcome. Coal and petroleum companies, for example, are formidable barriers to shifts away from the use of fossil fuels in energy production. Conservation and other environmentally sound energy developments are also cost effective and will benefit economies as a whole, but the distribution of benefits and losses will be uneven and will generate powerful political pressures to avoid or minimize those losses. The distribution of the benefits and costs of regulatory alternatives and substitutes; the relationship among pollution-producing activities, poverty, and economic development; and the likely impact of regulatory alternatives and substitutes on economic development and on poverty, particularly in developing countries, all influence the development of policy responses to global environmental problems.

INCENTIVES FOR LESS-DEVELOPED COUNTRIES

Many people in the wealthy world debate whether there is a global environmental crisis, but people in the less developed countries do not have that luxury. Residents of the wealthy nations may have the resources to adapt to deteriorating environments, but those in the less-developed world will largely be unable to escape the consequences of global environmental change. Indeed, they have already suffered the consequences. An optimistic view holds that we have some years to decide how to respond to threats such as global climate change. But the gap between population growth and the ability to grow food because of land and water degradation, for example, is

Challenges in Implementing Environmental Agreements

already here and is growing.[41] Financial and technological assistance from the industrialized states to the developing countries will be essential if global agreements are to be implemented effectively.

The industrialized nations have led the debate over global environmental protection. They have faced more than others the consequences of industrial and agricultural pollution, research on environmental problems and consequences is largely done in developed countries; they have the resources available to go beyond survival to environmental protection; and they have well-organized environmental groups that demand action. The decline of the threat posed by the former U.S.S.R. has permitted countries to focus on other concerns.

The less-developed world has been less committed to global environmental regulation: Economic development has been seen as a more pressing need. These countries fear environmental protection efforts will divert money and attention from development needs, and many see pollution as the price of modernization and development. They have called for a new international economic order that recognizes underdevelopment, and not population growth, as the primary problem in less-developed countries. The 1980s resulted in economic decline and falling standards of living throughout Africa and in some other regions. The enormous foreign debts that burden many countries are being paid off through consumption of natural resources such as fossil fuels and deforestation, which also exacerbate global environmental threats.[42] (This discussion is pursued in later chapters).

Much of the challenge results from the distribution of environmental problems. Delegates from Malaysia, for example, fought against provisions proposed during the UNCED process that would prevent their nation from exploiting its rain forests. They vigorously opposed efforts aimed at an agreement to protect forests, arguing that the United States and others had largely harvested their forests and now demanded that the countries where forests remain be forced to preserve them as a sink for the carbon dioxide emissions from the industrialized countries. The poor island countries of the Pacific, in contrast, demanded that wealthy nations reduce their carbon dioxide emissions to decrease the threat of global warming and a rise in sea level that would inundate them.

Since both the North (most developed) and South (less developed) must participate in solutions to problems such as global climate change, environmental protection must be linked with development. The South has considerable leverage, given its enormous population. The North cannot ignore the demands of the South for economic growth and increased consumption. Equity will be a key consideration, and will be pursued through technology transfer, financial assistance, training and education, and so on. The potential is great for solving the environmental problems as well as the development concerns that have languished for decades.[43]

Creating Incentives for Compliance

Global agreements must provide for some kind of enforcement mechanism. One possibility would be to create or authorize an international organization to monitor compliance and issue sanctions. In a more decentralized approach, participating countries could take nations they accuse of violating agreements to an international tribunal for resolution of their disagreements. Some studies of global environmental agreements have focused on the importance of review mechanisms as the key to effective implementation, and, in particular, how alternative ways of reviewing compliance with international agreements might affect the level of implementation. Self-reporting of data is the most common method, but this raises questions about the accuracy of the data and the possibilities for developing more effective reporting systems. Some environmental problems are continuous, and reporting of compliance is an ongoing concern, whereas other problems can be solved immediately by specific, relatively short-term actions.[44]

THE LEVELS OF COMPLEXITY

One of the most important challenges in implementation involves the complex interaction of various global actors and levels of

Challenges in Implementing Environmental Agreements

government. Although global agreements are aimed at national governments, environmental problems are largely caused by transnational or multinational corporations (MNCs). How can states ensure the participation of transnational corporations in meeting global obligations? Besides being the major source of pollution, MNC's are also the major producers of pollution-control innovations. They play a critical role in the development and application of technological innovations to reduce pollution. MNCs dominate domestic politics and policies. They help create and alter consumer tastes and demands for products and services.[45]

A related challenge is matching the institutional, political, regulatory scale to the scale of pollution. At what level of government should we pursue different problems and implement different solutions? Lower levels of authority—local governments—are more competent in some areas than others, but they face their own shortcomings of limited expertise, jurisdiction, and resources. Environmental regulation is further complicated by interactions between central and subnational governments. Environmental regulation in the United States, for example, has evolved into an intricate system of shared authority and cooperative agreements between the federal government and the states, largely in response to the size of the nation, the complexity of environmental programs, and the tremendous number of sources of pollution to be regulated. Most federal environmental statutes authorize states to issue permits and to enforce regulations if their programs and standards are approved by the EPA. States have the primary responsibility to grant permits, inspect facilities, and initiate enforcement actions against violators. The federal government's primary function is to establish overall policy, develop national standards, and ensure that states enforce the laws and regulations in a way consonant with national standards.

The U.S. experience in environmental regulation is so much a function of unique political and institutional arrangements that differences from other countries are dramatic. Environmental activists in other countries, perhaps most visibly in West Germany, may gaze fondly at the access and influence of their U.S. counterparts and demand more participation in government and more confrontation with industry, but the regulatory experience of the 1970s is much more of a model for them than the regulatory relief of the 1980s. Recent administration attempts to "reform" regulation, with its emphases on considerations of costs, benefits, and industry-government cooperation, appear on the surface somewhat similar to the traditional European approach. But the adversarial nature of environmental regulation has flourished in the United States through aggressive congressional supervision, the enactment of increasingly specific and detailed statutes, active monitoring of regulatory activity by public interest groups and the media, and legislative lobbying and litigation by advocates of increased regulation. Although an attempt has been made to develop a more cooperative business-government relationship, skepticism of bureaucratic authority and expertise, fear of business-government collusion, and lack of consensus over scientific and economic analyses are resilient characteristics of the regulatory process in the United States.[46]

Political leadership can play another role in addition to formulating policies for energy development and environmental protection. Symbolic acts by political leaders, such as driving less, planting trees, or installing energy-efficient light bulbs, can focus attention on what ordinary citizens can do themselves to resolve global problems. Although governments formulate major policy interventions needed to make significant reductions from major sources of pollution, individuals acting in an environmentally responsible way can do much. Grassroots involvement is essential in educating the public about global environmental problems and the steps they can take to address these problems in their own lives. Individuals will be much more responsive if they understand why changes in behavior and sacrifices are necessary and if they have some freedom to choose those actions they wish to take on a personal level.

GLOBAL TREATY REQUIREMENTS

Few countries will be willing to impose stringent regulatory controls on their industries without assurance that competing industries in

Challenges in Implementing Environmental Agreements

other nations face similar requirements. The wide range of regulatory systems in different nations will increasingly be expected to create common environmental mandates. Global agreements, or at least accompanying documents, should include plans to assess on a regular basis how well participating nations are complying with agreements and what they are learning that can be exchanged with other nations.

What kinds of treaty provisions are most likely to elicit compliance? It is difficult to determine the extent to which particular treaty provisions cause specific actions. The interest of countries and the actions required by agreements may correlate with provisions of treaties and compliance, but there may be no clear causal links. The most significant impact of an agreement may be to set in motion subsequent efforts rather than specific responses. Treaties may combine with a host of economic, political, and other factors to shape behavior. Even the most ardent proponents of international law recognize that agreements are not always kept, particularly when they conflict with other important, pressing interests.

How can agreements be designed to foster compliance? Ronald Mitchell's study of compliance with treaties governing international oil pollution found several situations where treaty provisions and compliance systems appeared to produce behavioral change. Despite considerable noncompliance with treaty provisions, Mitchell found "extraordinary variance between levels of compliance across the provisions of a single treaty": different kinds of treaty rules and provisions that were aimed at the same behavior produced "strikingly different levels of compliance."[47] Compliance systems include the primary rules and regulations, information on the level of compliance, and the response to evidence of noncompliance. Agreements that were successfully implemented ensured that the tasks of compliance, monitoring, and enforcement "placed the burden of performing those tasks on actors that have the political and economic incentives, the practicable ability, and the legal authority to carry them out."[48]

Two conclusions are particularly important. First, the kinds of prohibitions and prescriptions included in treaties and the system of monitoring and responding to violations have a significant impact on the extent of compliance. Rules that "increase compliance more by matching burdens and responsibilities with actors' interests and capabilities" are more likely to be successful than those that seek to alter those interests and capabilities; "successful rules have facilitated and removed barriers to desirable action by those actors with existing interests in taking those actions." Second, information-collection systems should be designed to provide benefits to the countries providing the information: "whether they involve self-reporting or independent verification, the compliance information systems that have succeeded in eliciting compliance data have done so by making use of the information collected to further the interests of those providing it" and by building on "preexisting information structures."[49]

Mitchell and others argue that the key elements of a compliance system are the primary rules, the compliance information mechanisms, and noncompliance response mechanisms. The primary rules are the pressures and incentives to achieve compliance with the agreement's provisions. The compliance information mechanisms are the reporting, monitoring, data analysis, and dissemination activities. The noncompliance mechanisms include the responses available in cases of noncompliance. Unlike some other areas of international law, compliance with international environmental agreements requires that national governments devise regulatory schemes to change the behavior of subnational governmental and private actors, rather than devising rules to govern the behavior of the state itself. The primary rules determine which interests are to be regulated and who must change their behavior. Compliance is to a great extent a function of such factors as the kind of activity to be regulated, the number and diversity of actors, and the magnitude and the cost of the required changes. The transparency of compliance activities is also important: The more visible the actions of the regulated parties and the more information available concerning compliance, the more likely the other parties will be reassured that compliance is occurring. Different kinds of rules may influence the likelihood that regulated parties will comply. Rules that include

Challenges in Implementing Environmental Agreements

specific, measurable performance goals may increase compliance as they make clear the obligations imposed on regulated parties. However, as rules become more specific, agreement may become more difficult as parties have less room to compromise around vague language.[50]

Transparency is also critical to efforts to produce compliance information. The kind of primary rules selected affect transparency: It is easier to verify compliance with some regulations, such as production prohibitions, than are other regulations. Some compliance-related information may already be collected, while the creation of new monitoring and reporting efforts may still be needed. The self-reporting that most treaties require can be supplemented through inspections and on-site monitoring. Agreements that provide for the means to collect, analyze, and disseminate compliance information can create important incentives for compliance by giving parties the information necessary to invoke sanctions, pursue reciprocal relationships, and apply incentives.[51]

Once noncompliance is discovered, several course of action can change behavior and induce compliance. Funding can be provided to help finance compliance efforts. Educational efforts can encourage the motivation to comply and provide technical assistance. Transfer of necessary technology can occur. Incentives may be more politically feasible than sanctions, but they may be unnecessary since some parties that receive them might have complied even in the absence of incentives. International meetings can provide a forum for identifying noncomplying parties and pressuring them to change. Treaties can specify the sanctions to be applied when noncompliance is uncovered. Incentives can be created to encourage monitoring by private actors, including competitors of regulated industries. Agreements may permit governments to take enforcement actions against the citizens of other nations found violating provisions. Inspection can also be encourage in ways that identify problems before noncompliance occurs, thus preventing violations. Central authorities, states, and private organizations can impose sanctions on nations, industries, local governments, and individuals. Sanctions can take the form of trade restrictions, cancellation of contracts, consumer boycotts of products, media-based campaigns to change behavior, and other efforts aimed at influencing governments through public opinion. More powerful states can compel compliance by others through providing resources or threatening sanctions. Strong states may feel they can ignore demands for compliance, whereas weaker states may have no choice but to comply even when their immediate self-interest dictates noncompliance.

Compliance may vary over time, as the relative power of states changes and their commitments and interests fluctuate. In the longer term, treaties can help build support for compliance through research and education efforts that change the perceptions of nations and help them see how compliance is in their self-interest. Dissemination of information can encourage domestic political forces to exert pressure on governments to meet international obligations. Treaty provisions can encourage states to give increased attention to environmental concerns.[52]

In sum, Mitchell's study of global environmental agreements argued that primary rules should:

1. ensure that governments and nongovernmental actors have the incentives, ability, and authority necessary to comply;
2. reduce monitoring burdens as much as possible; and
3. Frame obligations in specific terms that identify clearly who must do what.

Compliance information systems should:

1. reduce the resources needed to collect and report information;
2. consolidate and disseminate information in ways that are consonant with the regulated parties' own goals;
3. be as user friendly as possible; and
4. include independent verification mechanisms and remove legal and other barriers to inspections and other means of verification.

Challenges in Implementing Environmental Agreements

Noncompliance response systems should:

1. remove legal obstacles to the use of effective sanctions;
2. reduce the frequency with which responses to noncompliance must be made; and
3. ensure that responses are appropriate to the nature of the noncompliance.

The study concluded that treaty rules "independently influenced behavior" in some cases, but in other instances had little or no impact. "Treaty provisions have altered behavior when they have placed at least some actors within . . . an incentive-ability-authority triangle."[53] Effective compliance also requires altering the behavior of governmental and nongovernmental actors. International law, though itself difficult to enforce, can encourage traditional enforcement efforts when states enact legislation.

Finally, compliance and effectiveness are not the same thing. In some cases, compliance may be irrelevant; even low levels of compliance could lead to effective resolution of problems.[54] We need to evaluate how effectively the efforts launched by the agreement resolve the problem at issue. We must also devise consistent practices and standards while permitting some diversity in implementation, and balance the centralized review with diversity in implementation.[55] Learning is essential: International agreements must evolve along with our understanding of the problems we seek to remedy.

Promoting Compliance, Evaluation, and Adjustment

International agreements are ultimately dependent on enforcement. Some agreements carry moral force. Agreements concerning genocide, for example, are powerful because they are seen as moral statements. If the international community comes to view environmental problems as moral issues, it may be more willing to seek criminal, rather than economic-based enforcement and enforcement may be more likely to occur. The Montréal Protocol could be viewed as imposing moral obligations on states: "The Protocol, though acting in economic terms, reflects the greater moral value which the global community places on environmental quality, and, ultimately, on self-preservation."[56]

National and international regulatory schemes often establish judicial enforcement mechanisms to create incentives for compliance. Much of international environmental law relates present conduct to past decisions, such as claims concerning the health effects of exposure to toxic chemicals or nuclear waste or the environmental effects of air and water pollution. Nations have a general duty to inform and consult with other states when they become aware of possible transborder environmental damage.[57] International law also requires nations to inform others of actions that threaten the environment. States are to exercise "due diligence" to prevent a breach of international duty. This duty is to "refrain from acts that would cause injury to persons or property located in the territory of another state."[58]

Once a transboundary environmental violation occurs, the source state becomes responsible for remedying the problem even though the polluting facility might be a privately owned factory. Private companies rather than governments cause much pollution and natural resource damage. To invoke state responsibility in international arbitration, however, the state bringing the action must establish that (1) the offending state itself was responsible for the action; (2) there was a breach of international duty; (3) causal relationship exists between the action and the harm; and (4) material damage resulted. This mismatch between actual and legal responsibility for pollution poses a serious barrier to effective enforcement.

The World Court, or International Court of Justice (IJC), composed of fifteen judges who are elected by the UN General Assembly, provides a forum for dispute resolution. Unlike most American courts, the IJC is a court of specified jurisdiction. It also works on the principle that a state cannot be compelled to be a defendant without its consent. For example, in 1985, the United States withdrew its consent to be bound by the court's decision when Nicaragua brought action against it over a dispute concerning the mining of Nicaraguan waters by American forces. The court operates on an ad hoc basis as cases arise. In resolving disputes, the court relies on (1) international conventions that have established rules expressly recognized by the contesting states; (2) international custom, as evidence of a general practice accepted as law; (3) general principles of law recognized by civilized nations; and (4) judicial decisions and scholarly writings from various nations.[59]

The court has dealt with a wide range of environmental and natural resources issues, including the conservation of ocean resources, withdrawal of water, and the discharge of pollutants that affect international rivers and canals. International law recognizes the principle that states are liable for the "consequences of operations likely to cause serious harm to other states."[60] This principle might be applied to air and water pollution, but nations have thus far been unwilling to submit to decisions that might result in significant economic sanctions through the operation of the World Court.

However, court does not have compulsory jurisdiction. Only a few cases have been brought to international arbitration; no cases were brought for major transboundary pollution problems such as Chernobyl or the Sandoz chemical spill into the Rhine River. The cases brought are generally limited to narrow issues based on the wording of treaties, not on broad notions of liability. Standing is restrictively granted, and outside parties cannot monitor proceedings (only states, not individuals, have standing in international courts, and foreign citizens usually have a difficult time establishing standing in courts of other nations). In 1972, the Stockholm conference formulated Principle 22: States have "the responsibility to insure that activities within their jurisdiction or control do not cause damage to the environment of other States or of areas beyond the limits of national jurisdiction." But little further has been done to develop the law of liability and compensation. A number of questions remain unanswered: What level of damage is unacceptable? Should negligence or strict liability be the standard used in dispute resolution? Negligence is the standard in many nations, but it is interpreted differently in different nations. The uncertainty surrounding international environmental law limits the power of international arbitration.[61]

Critics also argue that this reactive approach to environmental protection encourages confrontation and discourages cooperation. It fails to prevent problems or provide incentives to reduce pollution. It does not ensure that the benefits of participation exceed the benefits of noncompliance. The legal approach is based on abstract norms of states' rights that fail to account for a common agenda of states' interests. States view submission to the authority of these institutions as a violation of their sovereignty, which is "immutable and a most sacred principle of international law."[62]

The experience of Europe in dealing with Chernobyl and acid rain demonstrates other significant barriers to effective environmental law. These barriers include global economic competitiveness, limited opportunities to resolve conflicts, and uncertainty surrounding the causes and consequences of certain global environmental problems. Some reforms have been proposed, including compensating individuals by states for damage done by sources within their borders; requiring insurance to ensure compensation for pollution victims; recognizing pollution as an international crime; and recognizing a clean and healthy environment as a human right.[63]

Despite the shortcomings in environmental law, international law has been rather successful in regulating economic, commercial, technical, and social activities such as international telecommunications, and in controlling health epidemics. Most states are willing to comply voluntarily with these restrictions on their sovereignty as long as others make the same concessions. They widely perceive cooperation to be in their own interest and agree to comply with international agreements on these subjects, to incorporate their provisions into national law, and to use domestic enforcement mechanisms to ensure compliance with the law. Many treaties also provide for the monitoring and enforcement of agreements through organizations within the UN structure or through independent agencies such as the International Atomic Energy Agency. International law relies heavily on the willingness of states to subject their performance to international scrutiny. Dispute resolution efforts include arbitration, conciliation, mediation, and negotiation. Sanctions include economic and trade limits, political pressure, and occasionally, as in the case of the 1990 Iraqi invasion of Kuwait, military action.[64]

Environmental protection, however, has less a tradition of global cooperation. States have been less willing to give up a little of their sovereignty in the environmental interests of the international community, because they regard the economic interests that are affected as so significant. States need to recognize that they must sacrifice some sovereignty to ensure global environmental and economic security, that national self-interest lies in protecting the global commons. But sovereignty will probably continue to be a major barrier, so international efforts that do not threaten sovereignty will be required. It is likely that international law will "only be auxiliary to the political, technical and psychological aspects of a solution" to major global environmental problems.[65]

Prospects for Effective Implementation

The complexity of the problems to be addressed and the interdependence of so many factors places tremendous stress on our governing institutions. Administrative fragmentation, the separation of powers, and other characteristics of government are imposing challenges in the United States and many other nations. Bureaucratic specialization and division of labor conflict with the holistic, comprehensive approach that the nature of the biosphere requires.[66] As this book frequently argues, solutions to these problems must be implemented at all levels of society. Although national governments are the primary focus of attention in global accords, the real focus must be on the actions of industries, consumers, local governments, communities, and individuals. The efforts of international bodies, multinational corporations, nongovernmental scientific and environmental groups, and many others must be integrated with official state action. Successful implementation will result from a great diversity of disparate actions that can somehow be shaped around common purposes.

The breadth of global environmental threats and their potential impact, especially on those who lack mitigating resources, require that we take precautionary steps. Incentives for dramatic changes in behaviors are needed. But even if we understood most problems well enough to launch comprehensive remedies, our political systems are oriented toward slow, incremental changes in the absence of unambiguous crises. We need to build on the strengths of an incremental approach to policy making and at the same ensure that we are moving in the right direction. We need to learn how to more effectively aggregate our experience, learn from mistakes, and make midcourse corrections.

References

1. Thomas Kamm, "Some Big Problems Await World Leaders At the Earth Summit," *Wall Street Journal*, 29 May 1992, A1.
2. Steve Fainaru, "Down to Earth," *Boston Globe*, 7 June 1992, 81.
3. For a sampling of this literature, see, generally, Thomas E. Mann, ed., *A Question of Balance: The President, The Congress and Foreign Policy* (Washington, D.C.: Brookings Institution, 1990); Robert E. Hunter, Wayne L. Berman, and John F. Kennedy, *Making Government Work: From White House to Congress* (Boulder, Colo.: Westview Press, 1986); National Academy of Public Administration, *Beyond Distrust: Building Bridges Between Congress and the Executive* (Washington, D.C.: NAPA, 1992); and John E. Chubb and Paul E. Peterson, eds., *Can the Government Govern?* (Washington, D.C.: Brookings Institution, 1989).
4. Books that raise many of these issues include Al Gore, *Earth in the Balance* (Boston: Houghton Mifflin, 1992); Jessica Tuchman Matthews, ed., *Preserving the Global Environment* (New York: Norton, 1991); Peter M. Haas, Robert O. Keohane, and Marc A. Levy, *Institutions for the Earth* (Cambridge, Mass.: MIT Press, 1993); and Sheldon Kamieniecki, ed., *Environmental Politics in the International Arena* (Albany, N.Y.: State University of New York Press, 1993).

References

5. See John W. Kingdon, *Agendas, Alternatives, and Public Policies*, 2nd ed. (New York: HarperCollins, 1995).
6. For a classic exposition on the nature of the policy-making process, see Charles O. Jones, *An Introduction to the Study of Public Policy*, 3rd ed. (Monterey, Calif.: Brooks Cole, 1984).
7. Charles E. Lindblom, *The Policy-Making Process* (Englewood Cliffs, N.J.: Prentice-Hall, 1968), 4.
8. Jeffrey Pressman and Aaron Wildavsky, *Implementation* (Berkeley, Calif.: University of California Press, 1984), xxi–xxiii.
9. Charles E. Lindblom and Edward J. Woodhouse, *The Policy-Making Process*, 3rd ed. (Englewood Cliffs, N.J.: Prentice-Hall, 1993), 10–11.
10. For an exploration of these and other themes, see Daniel A. Mazmanian and Paul A. Sabatier, *Implementation and Public Policy* (Glenview, Il.: Scott, Foresman, 1983).
11. See, generally, Eugene Bardach, *The Implementation Game* (Cambridge, Mass.: MIT Press, 1977); Andrew Dunsire, *Implementation in a Bureaucracy* (Oxford: Martin Robinson, 1978); Malcolm Goggin, Ann Bowman, James P. Lester, and Peter O'Toole, Jr., *Implementation Theory and Practice* (Glenview, Ill.: Scott Foresman/Little Brown, 1990); Daniel Mazmanian and Paul Sabatier, eds., *Effective Policy Implementation* (Lexington, Mass.: Lexington Books, 1981); and Robert T. Nakamura and Frank Smallwood, *The Politics of Policy Implementation* (New York: St. Martin's, 1980).
12. Pressman and Wildavsky, *Implementation*, 176, 180.
13. Ibid., 180.
14. Ibid., 168–80.
15. See Theodore Lowi, *The End of Liberalism*, 2nd ed. (New York: Norton, 1979).
16. See, generally, Burke Marshall, ed., *A Workable Government? The Constitution After 200 Years* (New York: Norton, 1987); Donald L. Robinson, *Reforming American Government: The Bicentennial Papers of the Committee on the Constitutional System* (Boulder, Colo.: Westview Press, 1985); Hunter, Berman, and Kennedy, *Making Government Work*; James MacGregor Burns, *The Power to Lead: The Crisis of the American Presidency* (New York: Simon & Schuster, 1984); and James L. Sundquist, *Constitutional Reform and Effective Government* (Washington, D.C.: Brookings Institution, 1986).
17. Lowi, *End of Liberalism*, especially chapters 5 and 10.
18. Lindblom and Woodhouse, *The Policy-Making Process*.
19. Lindblom and Woodhouse, 13–22. See also Alice Rivlin, *Systematic Thinking for Social Action* (Washington, D.C.: Brookings Institution, 1971).

20. Lindblom and Woodhouse, *The Policy-Making Process*, 6–7.
21. Pressman and Wildavsky, *Implementation*, xix–xxv.
22. Mazmanian and Sabatier, *Effective Policy Implementation*, 3–35; see also their *Implementation and Public Policy*, 20–21.
23. Mazmanian and Sabatier, *Effective Policy Implementation*, 21–25.
24. Ibid., 25–30.
25. Pressman and Wildavsky, *Implementation*, 143–45.
26. Mazmanian and Sabatier, *Effective Policy Implementation*, 41.
27. International Institute for Applied Systems Analysis, "Implementation and Effectiveness of International Environmental Commitments" (January 1995).
28. A. O. Adede, "International Environmental Law from Stockholm to Rio—An Overview of Past Lessons and Future Challenges," *Environmental Policy and Law* (22/2 1982): 88–105, at 99.
29. World Resources Institute, *The Crucial Decade: The 1990s and the Global Environmental Challenge* (Washington, D.C.: WRI, 1989).
30. Michael Oppenheimer, "Context, Connection and Opportunity in Environmental Problem Solving," *Environment* 37 (June 1995): 10–15, 34–38, at 11–12.
31. See, for example, Gregg Easterbrook, *A Moment on the Earth: The Coming Age of Environmental Optimism* (New York: Viking, 1995); Ronald Bailey, ed., *The True State of the Planet* (New York: Free Press, 1995); and Wallace Kaufman, *No Turning Back: Dismantling the Fantasies of Environmental Thinking* (New York: Basic Books, 1994).
32. William Ophuls and A. Stephan Boyan, Jr., *Ecology and the Politics of Scarcity Revisited* (New York: W. H. Freeman, 1992), 195.
33. See Elinor Ostrom, *Governing the Commons: The Evolution of Institutions for Collective Action* (Cambridge: Cambridge University Press, 1990).
34. Nazli Choucri, "Introduction: Theoretical, Empirical, and Policy Perspectives," in *Global Accord: Environmental Challenges and International Responses*, ed. Nazli Choucri (Cambridge, Mass.: MIT Press, 1993), 3–4.
35. Peter M. Haas with Jan Sundgren, "Evolving International Environmental Law: Changing Practices of National Sovereignty," in *Global Accord: Environmental Challenges and International Responses*, ed. Nazli Choucri (Cambridge, Mass.: MIT Press, 1993), 401–29, at 419.
36. Ronald B. Mitchell, *International Oil Pollution at Sea: Environmental Policy and Treaty Compliance* (Cambridge, Mass.: MIT Press, 1994), 32–34.
37. Ibid., 34–37.
38. Ibid., 38–41.
39. Ibid., 35–36.

References

40. Ibid., 41–46.
41. See Lester R. Brown, "Facing Food Security," in *State of the World 1994*, ed. Lester R. Brown et al. (New York: Norton), 177–97.
42. Marvin S. Soroos, "From Stockholm to Rio: The Evolution of Global Environmental Governance," in *Environmental Policy in the 1990s*, ed. Norman J. Vig and Michael E. Kraft, (Washington, D.C.: CQ Press, 1994), 299–321, at 307–08.
43. Oran R. Young, "Negotiating an International Climate Regime: The Institutional Bargaining for Environmental Governance," *Global Accord: Environmental Challenges and International Responses*, 431–52.
44. Abraham Chayes and Antonio Chayes, "On Compliance," *International Organization* 47, no. 2 (spring 1993): 175–205; Mitchell, *International Oil Pollution at Sea*.
45. See Nazli Choucri, "Multinational Corporations and the Global Environment," in *Global Accord: Environmental Challenges and International Responses*, 205–53.
46. See, for example, David Vogel, *National Styles of Regulation: Environmental Policy in Great Britain and the United States* (Ithaca, N.Y.: Cornell University Press, 1985); Alan Peacock, ed., *The Regulation Game: How British and West German Companies Bargain with Government* (Oxford: Basil Blackwell, 1984); Charles E. Zielger, *Environmental Policy in the USSR* (Amherst, Mass.: University of Massachusetts Press, 1987); and Lester Ross, *Environmental Policy in China* (Bloomington: Indiana University Press, 1988).
47. Mitchell, *International Oil Pollution at Sea*, 10.
48. Ibid., 10.
49. Ibid., 12–13.
50. Ibid., 56–57; Patricia W. Birnie and Alan E. Boyle, *International Environmental Law and the Environment* (New York: Oxford University Press, 1992), 136–87.
51. Mitchell, *International Oil Pollution at Sea*, 57–59; Lynton Keith Caldwell, *International Environmental Policy*, 3rd ed. (Durham, N.C.: Duke University Press, 1996), 303–38
52. Mitchell, *International Oil Pollution at Sea*, 46–65.
53. Ibid., 312–27, 301–06.
54. See Oran Young, "Compliance and Public Authority," (Washington DC: Resources for the Future, 1979).
55. See James N. Rosenau and E.-O. Czempiel, *Governance Without Government: Order and Change in World Politics* (Cambridge, Mass.: Cambridge University Press, 1992).

56. Michael David Ehrenstein, "A Moralistic Approach to the Ozone Depletion Crisis," *University of Miami Inter-American Law Review* 21 (Summer 1990): 611–35, at 632.
57. See generally, Patricia W. Birnie and Alan E. Boyle, *International Law and the Environment* (Oxford: Clarendon Press, 1994); Alexander Kiss and Dinah Shelton, *International Environmental Law* (Ardsley-on-Hudson, N.Y.: Transnational Publishers, 1991; 1994 supplement); Edith Brown Weiss and Paul C. Szasz, *International Environmental Law: Basic Instruments and References* (Ardsley-on-Hudson, N.Y.: Transnational Publishers, 1992); Edith Brown Weiss, ed., *Environmental Change and International Law* (Tokyo: United Nations University Press, 1992); Alfred C. Aman, Jr., *Administrative Law in a Global Era* (Ithaca, N.Y.: Cornell University Press, 1992); Gerhard von Glahn, *Law Among Nations: An Introduction to Public International Law* (New York: Macmillan, 1992); and Daniel Patrick Moynihan, *On the Law of Nations* (Cambridge, Mass.: Harvard University Press, 1990).
58. Notes from Editors of the Harvard Law Review, *Trends in International Environmental Law* (Chicago: American Bar Association, 1992): 17–20.
59. Ian Brownlie, *Principles of Public International Law* (Oxford: Clarendon, 1979), 3.
60. Ibid., 285–86.
61. Notes from Editors of the Harvard Law Review, *Trends in International Environmental Law*, 33–34.
62. Ibid., 102.
63. Todd Howland, "Chernobyl and Acid Deposition: An Analysis of the Failure of European Cooperation to Protect the Shared Environment," *Temple International and Comparative Law Journal* 2 (Fall–Spring 1988): 1–37.
64. World Bank, *World Development Report 1992* 154 (Washington, D.C.: World Bank, 1992).
65. Oscar Schlachter, "The Emergence of International Environmental Law," *Journal of International Affairs* 44 (Winter 1991): 457–93, at 473–74.
66. Ophuls and Boyan, *Ecology and the Politics of Scarcity*, 248–49.

3

Alternatives for Achieving International Environmental Goals

MESA VERDE NATIONAL PARK in Southwest Colorado contains the ruins of the Anasazi people, who lived on the mesa for 1,300 years. They came to the mesa around the time of Christ, attracted by the fertile soil and increased rainfall. For some twelve centuries, the resources of the mesa sustained them. Scarcity of fuelwood, or perhaps other problems, caused them to begin to descend the mesa's cliffs and construct homes beneath the sandstone rock, where they were more sheltered from the elements. For nearly a hundred years, as builders of great skill and sophistication, they crafted villages out of carved rock. Further depletion of the soil, drought, deforestation, and other problems apparently forced them to abandoned their cliff dwellings at the end of the thirteenth century. They moved south, leaving their villages intact, to be discovered by white settlers in the late 1800s. The plight of the Anasazi is largely a mystery to archaeologists and anthropologists. The residents left few clear signs of what caused their flight. But the best available evidence indicates that they overconsumed scarce resources and were unable to adapt to a changing environment.

The history of the human population is replete with such stories. Refugees have been forced to flee droughts, floods, radioactive and toxic chemical releases, earthquakes, and windstorms. Indigenous peoples are often cited for their ability to adapt to various environmental conditions and to live in harmony with diverse ecosystems. However, at times their efforts have fallen short, and like the Anasazi, they have not been able to escape the consequences of ecological damage. Unlike native peoples, people in the technologically and industrially developed world largely believe that scientific progress has made them immune to nature's calamities, that natural threats have largely been conquered. One of the many paradoxes of modern societies is that although technology promises to insulate us from ecological forces, it also increases the magnitude of environmental trauma that does occur. As the global environmental predicament becomes more serious, more and more people will be increasingly pressured to abandon their homes for more hospitable environs. The resulting disruption will be enormous. But more significantly, the options for the planet's nearly 6 billion people who might try to escape environmental problems are quite limited. As the planet's population reaches 10 billion and beyond, and the problems are magnified, escape becomes even less and less an option.

Chapter 3 examines alternative roles for international institutions in implementing solutions to global environmental problems. There are significant limits to what these international institutions can do; alternative, decentralized mechanisms are also needed to create effective and efficient incentives to change behavior. What kinds of legal instruments are available to implement global environmental agreements as efficiently and effectively as possible?

Criticisms of Conventional Regulatory Approaches

Under the conventional, or command-and-control, approach to environmental regulation, governments issue national standards that usually take one of three forms: ambient standards that limit total concentrations of pollutants; emission standards that limit

Criticisms of Conventional Regulatory Approaches

what individual sources can emit; or design standards that require the use of particular pollution control equipment or production processes. Environmental laws and regulations are expected to create clear incentives for industry to develop and use newer, cleaner technologies. One of the success stories of environmental regulation has been the development of new control technologies such as the catalytic converter for motor vehicles.

The command-and-control approach to regulation has strengths and weaknesses. For example, standards are made binding on individual sources of pollution through required construction and operating permits. Standards are more stringent for new sources than for existing ones; this forces improvements in technology at lower cost than would completely rebuilding existing sources. Technology-based standards can force all sources to reduce emissions. Regulations are sometimes criticized for mandating design rather than performance standards, although it is difficult to make such distinctions in practice. Design standards (specifying a certain percentage of reduction in emissions, for example) may be based on estimates of what specific technologies can produce.

Although some criticism of regulation views judicial enforcement as relatively inefficient and ineffective, critics usually focus on bureaucratic or administrative problems. Implementing environmental laws is particularly problematic because of the breadth of the regulatory agencies' jurisdiction, scientific uncertainties concerning the causes and consequences of pollution, and political and ideological conflicts. Regulatory decision making, particularly the issuance of rules and technical guidance documents, is slow and cumbersome. The U.S. Environmental Protection Agency, for example, is often criticized for tardiness and for giving states and regulated industries little time to implement programs before compliance deadlines occur. Another complaint is that governments often fail to implement environmental laws, either because of a lack of expertise or because of a lack of political will to enforce laws and regulations effectively and to challenge economic powers and priorities. Although some of these problems are characteristic of governmental bureaucracies in general, or result from decisions made by other policy makers, they are nevertheless central to understanding the political context in which environmental laws are implemented.

A second category of complaints centers on the means employed to achieve environmental goals. Environmental regulations are inefficient, some argue, because they impose nationwide standards on sources and problems that differ according to local conditions. These regulations require all sources within the specified categories to comply, even when the costs of compliance and the benefits to be gained differ greatly. Design standards fail to give industries the flexibility they need to operate efficiently and effectively, rather than offering an opportunity to tailor their compliance. Standards mandate control technologies that lock industries into existing equipment and processes, and fail to create incentives for new, cleaner, more efficient technologies.

Finally, environmental regulations may fail to ensure that the benefits outweigh the costs. The cost effectiveness of regulations varies considerably, raising concerns about the expense of some regulations.

Much of the criticism of regulatory methods in the United States holds that the EPA and states have been slow to find alternatives to the command-and-control approach to regulation. Although some of the decisions concerning regulatory means are within the discretion of the agency, some are mandated by Congress; environmental statutes often include the specific steps by which the EPA is to pursue the objectives delegated to it. These criticisms also intertwine with those aimed at the goals of environmental protection policy and with broader criticisms of regulation in general.

Environmental regulation, like regulation in general, is disparaged as adversarial and contentious. U.S. regulatory agencies operate in a much more litigious context than their counterparts in Canada, Japan, and Western Europe. Regulation favors collective efforts rather than reliance on market forces. Regulation conflicts with rather than being integrated with economic and other policy goals to ensure that economic activities are environmentally sustainable. Regulators and policy makers have failed to engage the public and facilitate understanding of and debate on the threats we face and the choices available.

Criticisms of Conventional Regulatory Approaches

Many criticisms of regulation are unavoidable, given regulatory goals that seek to force changes in the way regulated parties conduct business. These goals increase the cost of production, distribution, and marketing. They require significant resources to ensure compliance. But regulation is an essential link between global accords and changes in economic, commercial, and household activities. One of the most important and promising policy debates surrounding regulation focuses on the cost of regulation and how its goals might be achieved at lower cost.

The U.S. General Accounting Office's 1992 report, prepared for the Clinton administration, identified several challenges confronting the EPA and concluded that the agency "has often been unable to meet statutory mandates and to implement plans for addressing pollution."[1] The EPA must also supplement traditional regulatory strategies with market-based incentives and other approaches that will permit industry to achieve environmental goals at a lower cost.[2] The Vice President's National Performance Review, released in 1993, focused on environmental management in the federal government. It similarly criticized the traditional pollution-control approach to regulation (as opposed to pollution prevention).[3]

Paul Portney brought together a number of scholars to examine the state of environmental regulation in a 1990 report.[4] According to one author, the federal and state agencies have only a "half-hearted commitment" to monitoring compliance with laws and regulations.[5] Far too little continuous compliance monitoring takes place. The EPA has focused on the installation of pollution control technology, relying heavily on self-monitoring by the sources. Other problems include infrequent auditing, lack of vigorous enforcement efforts to catch violators, infrequent use of self-monitoring reports to trigger investigations, and penalties that are too small to have much impact on industry finances.[6] There have been some improvements in developing a uniform policy and pursuing more criminal prosecutions.[7] But it is still necessary to develop better monitoring devices, amend laws to permit unannounced inspections, and strengthen the connection between monitoring and enforcement.[8]

Representatives of regulated industries regularly raise concerns over the inflexibility of regulation, the uncertainty that comes from a lengthy regulatory process, and the short amount of time they are given to comply with mandates. They often blame regulators (and congressional staff members) for lacking manufacturing experience to guide them as they impose restrictions on industrial activity. They seek more cooperative interaction with regulators so that the environment can be improved permitting firms to satisfy financial and other objectives. This frustration and criticism culminated in the election in 1994 of a Republican-controlled Congress. The House of Representatives passed a number of bills in early 1995 to give industries relief from regulatory mandates, and gave unprecedented access to industry lobbyists to draft legislation to make it more difficult for agencies to issue new regulations and to rewrite major environmental laws.

Command-and-control regulatory approaches under the Resource Conservation and Recovery Act (RCRA) illustrate well some of the challenges to making regulation more efficient. Stringent standards for land disposal of hazardous wastes have led to increased incineration of wastes that may actually elevate public-health risks through emissions of hazardous air pollutants.[9] Land-disposal regulations resulted in the closure of many disposal sites, thus leading to a shortage of disposal sites. As a consequence, the EPA has hesitated to classify as hazardous some wastes that exist in large quantities, fearing that there will be no place to dispose of them. Although the permitted sites are secure, great quantities of waste are not being regulated.[10] The U.S. General Accounting Office warned that the statute's mandate of deadlines for EPA issuance of operating permits but not for the closure of sites that fail to meet operating standards has prevented the agency from focusing its attention on sites—opened or closed—that pose the greatest environmental and health risks.[11]

The "regulatory reform" literature has long emphasized that more "efficient" regulatory strategies can replace current command-and-control approaches.[12] Critics argue that we cannot afford the inefficiency inherent in traditional regulatory schemes and that regulation often insulates some companies from competitive forces. National standards have been roundly rejected as inefficient and wasteful, since pollution problems are much more serious in some areas than others.[13]

Implementing Environmental Agreements Through Economic Incentives

Making regulation more efficient usually means shifting from the traditional approach of national, technology-based standards established by the EPA and enforced by federal and state officials to the use of economic instruments and incentives to prevent rather than treat pollution. So although there is little agreement over what environmental goals should be, once they are established, there is wide support for approaches to regulation that create market-based incentives.

Economic incentives have been increasingly heralded as the most efficient and effective mechanism for accomplishing environmental goals. Since the costs of controlling pollution vary so widely among firms, efficiency demands that "the degree by which individual sources have to reduce their pollution discharges should vary."[14] If sources of pollution are permitted to find the cheapest way to reduce emissions, the costs of pollution control will be minimized. Pollution is a failure of the market that can be remedied by extending the price system to incorporate the release of waste material.[15]

Incentives are championed as being essential to "harnessing the 'base' motive of material self-interest to promote the common good."[16] It is simply not effective to condemn polluters as being immoral or selfish, their advocates argue; clear incentives would encourage them to change their behavior, to ensure that they take actions consonant with the public good. Incentives also promote flexibility and freedom of choice and minimize the need for coercion. Putting a price on pollution can require firms to pay the full or true costs of what they produce, including the costs of diminished health or damage to property. This encourages pollution-control efforts up to the point where further controls would cost more than the value of the damage prevented. Revenue that is generated can be used to reimburse victims or prevent the adverse effects of pollution.[17]

The ultimate value of economic instruments may be in the opportunities they create to generate support for environmental regulation. Advocates of increased regulation can find common ground with those who champion industry autonomy and flexibility. The debate over whether environmental goals are too stringent is extremely contentious; reducing the cost of achieving those goals would soften some of that contention. If economic instruments can be devised to achieve environmental goals at a lower cost than the conventional regulatory approach, they will become one of the most important developments in regulatory policy.[18] Another important advantage of market instruments is that generally less information is needed by regulatory agencies. The burden is redirected toward industry engineers to devise ways to reduce emissions efficiently and in the least costly manner. The key question becomes much simpler: Did the company's emissions exceed the levels specified in its permit? The regulatory agency tasks are still considerable: the agency must estimate how much pollution exists in an airshed, distribute rights to pollute, determine which rights can be transferred, and monitor and penalize those who exceed limits. But these tasks may be less demanding than those under the command-and-control approach to regulation.[19]

In some cases, for example, administration of a regulatory program can be simplified by imposing a tax on inputs or materials rather than on effluents or emissions, since fewer firms may be subject to the tax and monitoring may be simpler. A charge on final products might be used in areas where input or emission charges are not effective or where environmental problems are primarily a function of consumer demand. This approach might create a more direct incentive for consumers to reduce consumption or switch to less damaging products.[20] Emissions charges create significant incentives for industries to reduce waste and save money. Such an approach requires careful monitoring and vigorous enforcement of requirements, but the revenues from the charges can finance these efforts. More difficult to overcome is industry opposition to having to pay for what it could do in the past for free, and the arguments that emissions fees and taxes would make it more difficult for industry to compete with foreign firms.[21] Market instruments will likely be used in tandem with traditional approaches, and may make implementation and compliance even more complicated. Public support for market incentives requires clear assurance that they will lead to reduced pollution as well as reduced compliance costs.

Several categories of marketlike mechanisms could be employed in environmental regulation: monetary incentives, including taxes, fees, subsidies, and tax incentives; government-created markets for trading emissions; deposit/refund systems that discourage disposal and encourage the collection or recycling of pollution-producing materials; disclosure of information to consumers; environmental auditing and the release of data on kinds and levels of emissions; and government procurement policies.[22] The two most controversial instruments are pollution charges and emissions trading.

Pollution Charges or Taxes

Under this approach, taxes are levied on emissions of pollutants or on inputs to activities that produce pollutants. Polluters are charged a fee for each unit of pollution they emit.[23] In theory, the tax should be high enough to generate resources that could compensate for the environmental damage resulting from the emissions. Such taxes ensure that marginal private costs, plus the added tax, equals marginal social costs. Revenue could be used to treat victims of air pollution, to pay for the damage to crops and buildings, to offset the cost of monitoring and regulating air pollution, or for other relevant efforts. A pollution tax or fee, if high enough, can provide a strong incentive for companies to reduce emissions in whatever way is most efficient for them—close down some operations, use cleaner fuels, invest in control technologies, change work practices, and so on. It also provides a clear incentive for reducing emissions below permitted levels. With the traditional approach to regulation, companies gain no economic advantage when they emit less pollution than is legally permitted them; with a pollution tax, they save money every time they reduce emissions. Pollution taxes, which operate like other cost factors in production, can preserve the flexibility and autonomous decision making that are important to businesses. Agency officials would also have a motive for enforcing the law if pollution tax revenues remain with the agency. Companies that comply with the law support strong enforcement efforts aimed at ensuring that their competitors also pay the required taxes.[24]

Tax incentives—deductions, exemptions, or credits—can be provided for particular types of investments. An accelerated deduction for depreciation of pollution-control equipment, for example, allows capital expenditures to be deducted for tax purposes well before their economic value is depleted. Alternatively, investment tax credits can be given to firms that invest in cleaner technologies. These incentives can help reduce the cost of complying with environmental regulations. Canada's experience with these incentives, however, "suggests that incentives are not always effective in achieving their goals." They complicate considerably the tax code and its administration and may "encourage more firms to enter an industry, thereby actually increasing pollution levels." They may also violate the "polluter pays principle" by spreading the cost to other taxpayers.[25] It is difficult to make the calculations required to ensure that taxes added to marginal private costs equal marginal social costs; an alternative approach would shift taxes away from labor and income to resources used in production, as a way to conserve them or reduce pollution.[26]

Emissions Trading

Under the emissions trading approach, polluters are allocated a limited number of allowances or units of emissions. These allowances, usually specified in operating permits that companies are required to have, require a reduction in pollution. Companies can either make the changes necessary to stay within their limits, or they can buy allowances from others. Emission permits can be sold, traded, banked, or saved for future use, as long as limits on total emissions are not exceeded. The process is relatively straightforward: Establish the ambient environmental quality goal (the maximum acceptable level of concentration of pollution); determine the total number of units of emissions required to meet the ambient goal (usually expressed in pounds or tons of pollutant during a specified time period); allocate permits to polluters to emit specified units of

Experiments with Market-based Regulatory Initiatives

pollutants by selling, auctioning, or giving them away; and enforce these limits by monitoring emissions from the regulated sources. Over time, total emissions can be reduced further by decreasing the number of allowances distributed to polluters.[27] Polluters have an incentive to reduce emissions even below their allowances so that they can generate revenues through the sale of excess allowances.[28]

Experiments with Market-based Regulatory Initiatives

Several market-based approaches to regulation have already been put in place in the United States.[29] Beginning in 1974, the EPA developed an emissions trading program that viewed emissions from each industry plant as encapsulated within a large "bubble," rather than from individual smokestacks. Regulatory officials established maximum total allowable emissions and then left managers free to determine optimal emissions from individual sources. Sources could "bank" emissions for future credits or sell them to new sources that needed to purchase offsets of existing emissions in order to operate.[30]

One of the most important innovations of the Clean Air Act of 1990 was the market-based incentive system to reduce acid rain producing emissions from coal-fired power plants. Placing a cap on total emissions is at the heart of the acid rain emissions trading system. By the year 2010, this will reduce sulfur dioxide emissions by 10 million tons from 1980 levels. The act requires 261 coal-fired power plants to begin reducing their emissions to about 2.5 pounds of sulfur dioxide for every million Btu of fuel they use beginning in 1995, and 1.2 pounds/million Btu beginning in the year 2000. Companies have several choices in achieving these reductions: retrofit their facilities with devices that remove SO_2 from their emissions, burn lower-sulfur coal, install new technologies that release fewer emissions, or purchase allowances from other companies. The EPA allocates annual emission limits for each facility, based on a formula Congress provided in the law. At the end of each year, sources must show that their emissions are no greater

ALTERNATIVES FOR ACHIEVING INTERNATIONAL GOALS

than their allowances; if emissions are greater, sources can buy or trade allowances from others. If they fail to meet their emission limits, they are fined for each excess ton and must reduce that amount from their allowable emissions in subsequent years.

The acid rain provisions are the most prominent use of economic incentives, but a similar approach is taken elsewhere in the 1990 Clean Air Act. Marketable permits are used to phase out the manufacturing of chemicals that deplete the ozone layer. Companies that reduce their emissions of toxic chemicals by 90 percent or more before the EPA issues emission standards are exempted from other controls. Emission fees for severe ozone nonattainment areas are provided for, and states are authorized to employ emission fees, auctions of emission rights, and other innovations in their regulatory programs.[31]

Several states have begun to develop plans for emissions trading programs.[32] The Illinois Environmental Protection Agency, for example, concluded in a September 1993 report that market-based approaches to reducing ozone levels would be less expensive than conventional regulation.[33] Illinois also instituted a pilot program of buying and scrapping pre-1980 autos. The state purchases the vehicles for $600 to $1,000, then measures the tailpipe emissions and fuel evaporation before destroying them.[34] The Northeast States for Coordinated Air Use Management, comprising the eight northeastern states, has proposed a regional trading program that promises to reduce the cost of achieving ozone standards.[35] Maryland is the first state to institute a "gas guzzler" law, which imposes a surcharge on the sale of cars that consume high levels of fuel and offers a rebate to buyers of fuel-efficient autos.[36] A consortium of New England industries has outlined plans for an emissions bank that would permit firms to save surplus credits and trade them in or sell them as they construct new facilities.[37] The Texas Air Control Board considered a range of programs, including emission reduction credit banking, community banking, early vehicle retirement, conversion of motor vehicles to alternative fuels, and NO_x trading.[38]

The most ambitious emissions trading scheme has been proposed in southern California. The South Coast Air Quality Management

Experiments with Market-based Regulatory Initiatives

District, the agency responsible for addressing the Los Angeles area's air pollution problems, developed an emissions trading program called the Regional Clean Air Incentives Market (RECLAIM).[39] This program, approved in 1994, promises to reduce hydrocarbon or reactive organic compound (ROC) emissions by 85 percent and nitrogen oxide emissions by 95 percent in order to bring the area into compliance with the national ambient air quality standard for ozone. (The 1990 Clean Air Act classifies the South Coast Air Basin as a severe nonattainment area for ozone.) Under this plan, oil refiners and other large industrial polluters are to reduce emissions by a fixed percentage each year for different pollutants. Companies that reduce pollution by more than the required amount can sell pollution credits to other companies that exceed their limits.[40] Sources are also required to reduce emissions by 5 percent a year from their baseline. Stationary sources may, under certain conditions, obtain credits from mobile sources by purchasing and retiring older model cars to extend compliance deadlines or increasing average vehicle ridership among employees.[41] In 1995, the SCAQMD reported that Los Angeles had made significant improvements in air quality. In 1960, ozone levels exceeded the national standard four out of every five days; in 1995, the standards were exceeded only every two of five days. These reductions were achieved despite a three-fold increase in motor vehicle traffic and population and the construction of new industries throughout the air basin. Los Angeles continues to have the worst air pollution in the nation, and its residents suffer more adverse effects than anyone else. But some officials even began expressing optimism that national air quality standards could be met by 2010.[42] The area has become a model of regional cooperation, and other states have adopted many of the rules to meet their own attainment goals.

In response to the dumping of hazardous wastes in unapproved sites to avoid the costs of proper treatment, the economist Clifford Russell has proposed a positive incentive system. Under his approach, generators and transporters would be paid when they brought toxic wastes to approved disposal sites. The challenge lies in setting the level of award. Any use of public funds for such rewards means that either taxes will have to be increased, thereby offsetting spending cuts made elsewhere, or deficits will have to be increased. The scheme could include a deposit-return component, similar to those developed for recycling beverage containers. When specified materials are purchased, a deposit fee is added to the price. The refund goes to whoever returns the material, creating an incentive for other parties to get involved in the waste recovery effort. In European countries, the deposit-return system has included, in addition to beverage bottles and cans, automobile parts, motor vehicle batteries, and lubricants.[43]

Perhaps more difficult, the incentive must be great enough to encourage compliance but not so great that companies are encouraged to produce the waste in order to receive the reward or dilute waste with other material in order to increase volume. One option is to require that purchasers of regulated materials be given a manifest, indicating the quantity purchased, which must be presented when returning the used materials for redemption. But this level of paperwork could be justified only for large purchases.

Identifying wastes is another major obstacle. It will likely be difficult to ensure that wastes discovered at a site are precisely those for which a refund is available, or that they have not been augmented by nonregulated wastes. Nor does this scheme deal with the problem of contaminants that dissipate into the environment and do not accumulate in recoverable quantities.[44] A deposit-refund program can provide effective encouragement for redemption and recycling. Policy makers will likely need to experiment with prices to find the level that will produce the desired result.[45] In that case as well, the key to policy success may be in policy maker's ability to learn from experience. Policy initiatives will need to include provisions for monitoring progress in achieving policy goals and making adjustments as needed.[46]

Waste minimization often requires significant capital investment, and these demands for capital must compete with others. It is expensive to gain the information required even to begin to solve the problem; identifying the components of each waste stream and determining the stream's source is a daunting and expensive task in many facilities. Changes in technologies can be expensive because they require investments in new and sometimes unproven technologies, involve shutdown and start-up costs, and affect product quality and plant operations.[47] Despite these difficulties, agreement among industry officials appears widespread that "waste reduction is a sound investment which makes good business sense." Companies have reduced waste generation by changing production techniques, using higher quality raw materials, developing more efficient processes, and finding markets to recycle materials that in the past would have been discarded.[48]

Another way to encourage the recycling of hazardous (and other) wastes is to place a tax on the use of virgin materials or to remove tax incentives and subsidies for harvesting virgin resources. Recycling can decrease the amount of hazardous materials that are discarded. For materials that are widely dissipated in use, a tax on the purchase of the materials themselves might be the most efficient and effective way to reduce consumption. A tax on the disposal of hazardous wastes on land can create incentives to produce less waste.[49]

An allowance system could also be established to limit the amount of waste that can be disposed of each year. Allowances to dump hazardous wastes could be issued to producers, who could buy, sell, and trade those allowances. No dumping would be permitted without the required allowances. The number of allowances could be decreased over time, thus giving producers a further incentive to reduce emissions beyond their allowances.[50]

Other challenges include how to reduce pollution generated by small sources and those not easily identifiable. It is not clear that a tax on small sources would be any easier to implement than the conventional regulatory approach. One policy option would focus on land disposal that poses long-term risks. A tax or allowance scheme could be imposed on wastes that, even when disposed of properly, are likely to pose a long-term risk to human health and the environment. The EPA would specify which wastes would be so classified, forcing the agency to set priorities and to determine which risks pose the greatest threats. Waste generators would have an incentive to demonstrate that their wastes are not a long-term threat; the burden would be on industry rather than the EPA to show that the wastes at issue should not be regulated as long-term hazards.[51]

Instead of the conventional approach, agencies could offer tax incentives to those who purchase sites to be cleaned up. Tax credits would allow parties could deduct from the taxes they owed, dollar for dollar, the amount they spent in cleanup. The regulatory agency would continue to be responsible for establishing the elements of the cleanup plan and ensuring compliance with it. However, although tax credits would decentralize cleanup efforts, they may not necessarily make cleanup more cost effective, since parties would have no incentive to reduce costs but could simply pass them through to the federal treasury. The regulatory agency (or perhaps the tax-collecting authority) would have to monitor cleanup activities closely to ensure that only reasonable costs are charged. A less expensive plan, from a government budget perspective, would be to grant tax deductions, rather than full tax credits, for remediation-related expenditures.[52]

The growing use in the United States of exactions imposed by local governments on developers who propose projects provides another option for a decentralized approach to dealing with hazardous waste sites. Developers have been required to dedicate lands or make case contributions to a fund for streets; sewer, water, and other utility lines; and schools and parks.[53] Some cities have required commercial office space developers to build low-income housing as a condition for receiving the necessary permits.[54] One proposal would permit local governments to require developers to clean up abandoned hazardous waste sites in order to gain a permit to construct a commercial or residential development. If the development would likely contribute significant amounts of waste to the municipal landfill, for example, the developer could be required to clean up an existing dump. Or developers could build on contaminated land if they agreed to clean up the site.[55]

Challenges in Using Market Instruments

The EPA's Economic Incentive Program (EIP) Rule, issued in April 1994, outlines some of the challenges of employing market instruments in environmental regulation programs. The rule requires state EIPs to include the following elements:

Clearly defined goals and an incentive mechanism that can be rationally related to accomplishing the goals;

A clearly defined scope that identifies affected sources and ensures that the program will not interfere with any other applicable federal regulatory requirements;

A program baseline from which projected program results (e.g., quantifiable emissions reductions) can be determined;

A requirement that reductions used to generate credits must be surplus, quantifiable, permanent, and enforceable;

Credible, workable, replicable procedures for quantifying emissions;

Source requirements consistent with specified quantification procedures and that allow for compliance certification and enforcement;

Requirements for projecting program results and dealing with uncertainty; and

An implementation schedule, administrative system, and enforcement provisions adequate for ensuring federal and state enforceability of the program.[56]

It can be difficult satisfy these standards when formulating and implementing regulatory programs that use market instruments. Although the theory of market instruments in regulation seems compelling, actually using these instruments raise a host of questions and concerns. Despite the simplicity of the market, proposals for market instruments are nevertheless often complex; discussion over the details of trading plans may obscure broader questions about how much reduction in emissions is required and how much should be spent to improve environmental quality. Poorly designed or misdirected regulatory goals would not be improved if the cost of complying with them was reduced through trading programs. Trading schemes may become the focus of attention themselves as interests argue over the specific provisions and mechanisms to be used. The environmental goals may be slighted or ignored, with discussions and analyses centering on how the trading will work rather than on whether or not it will reduce emissions. Emissions taxes and trading policies do not obviate the need for careful analysis of the goals of environmental policies. Even though market instruments can be applied to some policies, those policies may not make environmental sense or otherwise be worth pursuing.

Economic instruments can help improve environmental quality by internalizing the cost of environmental consequences of producing goods and services so that costs of production, and eventually the prices charged, reflect true costs. Despite the significant advantages that emissions charges represent in environmental regulation, they have not been adopted at levels sufficient to internalize costs. This is partly because it is difficult to calculate the levels of charges required to reduce emissions by the amount required to meet air quality goals. More important, as noted above, it is usually exceedingly difficult to overcome the opposition of regulated industries to new fees or charges. The Clinton administration's ill-fated effort in 1993 to impose a Btu tax on energy sources graphically describes the kind of political opposition that such proposals generate. Regulated industries support emissions trading programs and other mechanisms that give them increased flexibility and reduce their costs, but they obviously do not endorse regulatory approaches that increase their costs.

Shifting to a true-cost economy that relies on realistic prices is the best way to limit demand and to ensure optimal use of resources.[57] However, economic incentive innovations that stop with trading schemes do not sufficiently encourage the spread of the concept of true costs.

Challenges in Using Market Instruments

Emissions trading programs can be most beneficial in helping prepare the way for emissions taxes and fees by providing the institutional infrastructure as well as by reshaping expectations toward more effective means of regulating pollution. In the aggregate, trading schemes may internalize costs for an entire sector of the economy, but they do not require every source to internalize its costs. Companies that find it more expensive to clean up than to purchase emission reduction credits from others may do little to reduce their pollution. Community members who insist that major sources of pollution do all they can (while remaining economically viable) to reduce their emissions may be dissatisfied when those sources can escape that obligation. In one sense, trading schemes are also inconsistent with the "polluter pays" principle, which is one of the key ideas underlying environmental regulation. Trading tends to distribute equally the cost of pollution controls across all sources, rather than imposing the greatest control costs on the sources that produce the greatest emissions. As a result, some firms are still able to externalize some of their costs of production to other sources rather than accounting for all their costs.[58]

Environmental policy has aimed to encourage or force the development of cleaner, less polluting technologies. Despite flaws, the conventional approach to regulation has often expanded the use of cleaner technologies and encouraged the development of new technologies. This momentum can be lost if emission credits are cheaper than investing in newer, more environmentally friendly technologies. In contrast, emissions taxes provide a continuous incentive to devise new processes and technologies, because every time reductions are made, lower taxes result. Environmental regulatory efforts are in fact partly assessed by their technology forcing; emissions trading programs, unless aggressively structured, may reduce the pressure for those developments.

Challenges in Emissions Trading Programs

Since emissions trading programs have become the most commonly discussed market instrument for environmental regulation, the challenges they pose deserve particular attention. As outlined above, the process usually begins with a decision on the amount of emissions to be allowed. The next task is to calculate the volume of emissions that will not exceed the ambient standards. This requires an accurate inventory of existing emissions and the selection of the baseline to be used in allocating emission allowances. The selection of the baseline year is difficult, since emissions from sources of pollution vary considerably over time as a result of recessions and overall levels of economic activity, breakdowns and problems with maintenance and operations, investments in pollution-control equipment, and other factors. The initial allocation of emission credits is similarly critical: If it is too low or based on recession-year output, companies may not be able to comply when production increases. If the allocation is too high, real reductions may not occur for years. Permits can be allocated according to past and current emissions, or they can be auctioned to the highest bidder. The selection of a particular baseline year will benefit some firms over others. Firms that have already reduced their emissions may believe they are not being rewarded for being responsible when other sources that have done nothing are required to make the same level of reductions. Other questions include whether firms should be permitted to bank unused permits and use them later and how to deal with seasonal and other variations in emissions.[59]

A trading program requires a current, accurate emissions inventory for the calculation of a baseline or starting point. Many areas, however, lack the kind of inventory required to make the trading system work efficiently. Industry representatives who originally supported the marketable permit program have warned that the inventory needs to be updated and improved to account for fugitive emissions (those that cannot be traced to major smokestacks or point sources), incomplete data on some types of sources, and other shortcomings.[60] Emissions reductions that are to be traded should not be considered "surplus" until it is clear what levels of reductions are required to attain environmental goals.

Opportunities for regulated sources to gain emissions credits may inhibit the achievement of environmental goals. One of the most

Challenges in Using Market Instruments

attractive features of trading schemes, for example, is a declining cap that requires lower total emissions each year as a way to improve air quality gradually. However, loopholes can provide opportunities to circumvent the shrinking cap.[61] Emissions trading programs alone often do not necessarily improve environmental conditions, but must be combined with a cap on emissions that is ratcheted down over time, or emissions reduction credits that are discounted over time, to bring about a real reduction in pollution. Shrinking caps can be an important way of ensuring that trading schemes have environmental as well as economic benefits. Credits earned by plant shutdowns may create incentives for regulated industries to close existing facilities and move to new, less regulated areas, or to continue emissions that would have been reduced as plants close down for reasons unrelated to environmental regulation. Accurate past emissions from these sources may be difficult to obtain. Emissions credits from facilities that are shut down can be heavily discounted in response to these kinds of problems.[62]

Allowances can be sold, auctioned, or simply distributed to sources without charge. Charging for permits rather than distributing them freely has some important advantages. An auction can be a useful way to determine the value of the permit. The pricing mechanism is a means of automatically adjusting to market forces. This ensures the lowest expenditures for the required cleanup or the greatest amount of cleanup, given whatever we decide to spend on controls. Sources will undertake control efforts if they cost less than the price of the allowances. Some have even argued that markets should be extended to the number of allowances placed on the market, so that polluters and the community can negotiate what levels of pollution they want.[63]

Regulatory agencies can pursue more environmental protection than national standards require by decreasing the number of allowances, or they can permit environmental and other groups to buy allowances and reduce emissions even further. Sources would have a clear economic incentive to reduce emissions beyond their allotments and to sell their excess allowances to other sources. Both new and existing sources could be required to obtain allowances, thus ending the traditional approach of imposing more stringent standards on new sources than on existing ones. This may encourage the construction of new, less polluting sources, in contrast to the traditional approach that motivated companies to continue operating old sources and avoid the costs of complying with more stringent new source standards. Revenues from sold or auctioned allowances can be used to finance monitoring and enforcement efforts, to pay for research on pollution prevention, and to mitigate the effects of pollution. State regulators could also continue to rely on EPA issuance of technology-based standards, but permit sources to develop alternative technologies that meet or exceed the EPA mandate. This in turn reduces the burden on the EPA to update standards continually as new technologies are developed.

Trading programs can be combined with minimum technological controls that balance the flexibility from trading with the internalization of costs that comes from technology controls. Indeed, most emissions trading programs under clean air and other environmental laws dictate that those programs operate alongside technology control requirements, reasonable further progress requirements, measurable milestones, and other specific standards. The Clean Air Act's "expeditious attainment" requirement, for example, does not permit all reductions in emissions beyond the minimum, statutorily mandated requirements to be classified as surplus. Trading programs must first identify the most expeditiously attainable reductions possible, independent of any trading. Only true surpluses that represent more rapid or greater reductions than would occur under a nonconventional or nontrading regulatory scheme can be traded. As long as the regulated sources have different marginal costs, they will have an incentive to trade. The savings from trading must (1) be large enough to provide an incentive for trading; and (2) be used to reduce emissions in ways that are consistent with the statutory requirement that reductions be as expeditious as possible.[64]

Monitoring and enforcement are critical elements of the emissions trading scheme: in some ways, they are more important than in traditional regulatory schemes. Precise monitoring is essential. If compliance is not vigorously enforced, the incentive to clean up is lost. Standards that require sources to install pollution control equipment or change procedures are generally easier to monitor and enforce than emissions standards, since they require less sophisticated efforts. State officials may permit sources to monitor themselves, or they may assume that responsibility. This is particularly difficult in trading programs that cross state boundaries. The Northeast States for Coordinated Air Use Management and other multistate groups are exploring the possibility of an interstate emissions trading program for nitrogen oxides. It is not clear how enforcement of such a program would be integrated with state operating permits and implementation plans.[65]

Broader Challenges

The idea of buying permits to pollute appears to threaten the moral and symbolic power and appeal of pollution control; environmental regulation is no longer seen as a moral imperative in protecting human health—and, in particular, protecting innocent third parties who suffer few of the benefits of pollution-producing activity and most of the burdens—but simply as another cost of doing business. The most serious problem appears to be the distributional one: Air quality, for example, will improve for some people because some sources are cheaper to clean up than others. For those who live near facilities that are more expensive to clean up, however, emissions might increase.[66]

The acid rain emissions trading program has created distibutional problems: New York has discouraged utilities located within its boundaries from selling credits to companies operating upwind because of continued acid rain damage in the state. The amount of cleanup planned for in some areas is insufficient to prevent damage. Instead of enjoying reducing local emissions and improving local air quality, local residents pay higher prices but don't see leaner air. In contrast, a pollution tax would be a much simpler incentive for all sources to reduce emissions. The program has also permitted some emissions to increase: A Wisconsin utility company, for example, was permitted to sell 20,000 tons of credits to a Pennsylvania company, but Wisconsin law would have forbidden the utility from releasing those emissions in the first place.[67]

Reliance on pollution taxes and fees may also weaken public commitment to a shared environmental ethic. As decisions are left to the marketplace, educational and other collective efforts to foster public awareness of and concern for the environment may be viewed as less necessary. Although pollution taxes may be quite efficient, they may reduce the need for collective decision making about natural resources. If trading systems are used to create property rights in pollution emissions, for example, communities may have fewer opportunities to decide what level of air pollution they will accept. The kind of regulatory instrument selected may affect the political system and the policy-making process as well as environmental quality.[68]

Many advocates of free-market approaches to regulation tend to be quite selective in their embrace of economic incentives. Industries like emissions trading schemes that permit them to find cheaper ways to comply with standards, but oppose emissions fees that raise the price of doing business and other efforts to ensure that prices reflect more of the real costs of production. Market-based regulatory devices cannot ask the primary question regulators must answer: How much environmental protection is desired? But once that question is answered, economic incentives are promising ways to make regulation more efficient and effective, and to save money that can be used to pursue other public purposes. Market-based regulations work in areas where there are well-established markets, such as the production of commodities; raising prices to include all the costs of production will make markets better able to allocate scarce resources optimally. But economic incentives are not well suited to values that are not regularly priced through market exchanges. Many of the environmental values we cherish, such as protection of biodiversity, are not susceptible to market approaches. Biodiversity is valuable to us, for example, precisely because wild plants and animals and wild places have not become dominated by market exchanges. We limit markets in many other areas of life as well, even if there are willing buyers: Votes, judicial decisions, body parts, and a host of other valuable things cannot be bought or sold. Selling some things, such as spiritual gifts or intimacy or citizenship, actually degrades them. Markets are dominated by those with the most money; however, actions are not legitimized because they are paid for, but because they satisfy other values. Market-based regulation must be part of an effort to foster collective values and reinforce ethical commitments.[69]

Economic Incentives and Greenhouse Gases

The implementation of a global warming agreement poses particularly difficult challenges; because of the complexity, economic incentives may be the most efficient and effective way to reduce greenhouse gas emissions. The challenges involved in relying on economic incentives to implement a global climate change agreement illustrate the strengths and weaknesses of this approach to environmental regulation.

The number of sources of greenhouse gases is incalculable. The production of energy, the major source of greenhouse gases, is central to industrial and personal activity throughout the world. Efforts to check global warming interact in complicated ways with other environmental problems. There is considerable uncertainty concerning the relative contribution made by some greenhouse gases, the consequences should the climate get warmer, and the costs of preventive efforts. The consequences of global climate change and the costs of preventing it will not be equally distributed, but raise difficult issues of fairness and justice. Actions taken during the next few years may have irreversible consequences and major impacts on future generations. The costs of using fossil fuel should be borne now so that future generations will not inherit the burdens (without enjoying any of the benefits) of energy that is priced too low. Under any reasonable view of justice, the kind of mismatch between benefits and burdens that will result if warming projections are realized is simply not fair.

Given the costs involved in reducing greenhouse gas emissions, policy interventions must be particularly cost effective. Funds must be raised to assist emissions reduction efforts in poor countries. Catalysts to encourage the development of new technologies and pollution-reducing behavior must be developed. Sanctions and incentives will be necessary to ensure that participating nations achieve the emissions reductions required of them.

Another critical factor is the global distribution of greenhouse gas emissions (see Chapter 1). There are tremendous differences in the distribution of the sources of greenhouse emissions when measured in total and per capita emissions. Three countries—the United States, the former Soviet Union, and China—account for about one-half of global carbon emissions. Despite progress made in the last thirty years in the United States to reduce the amount of carbon emissions per unit of economic activity, Americans still produce more carbon per person than any other nation. Levels of carbon emissions are primarily a function of economic development; if the economies of the developing world expand significantly, the levels of greenhouse gases will likewise grow. For the present, many developing countries contribute to the growth of carbon dioxide levels in the atmosphere through their land use practices, particularly deforestation. Since population growth rates are so much greater in the developing world than the industrialized nations, carbon dioxide emissions from that part of the world are also expected to greatly expand. According to a 1990 report by the United Nations Population Fund, by the year 2025, Third World countries may be producing four times the amount of carbon dioxide they currently emit.[70] Pressure to improve the quality of life in these poor countries and their growing populations collide with concerns about the consequences of increased economic activity on the global environment. Unless that economic growth is radically different from current trends, the prospects for preventing global warming are bleak, even if the wealthy countries reduce their emissions.

For all these reasons, implementation of effective global climate change agreements may require market-based regulatory schemes. One option would require each country to make the same percentage reduction in greenhouse gas emissions. The simplicity of an across-the-board reduction, however, has some problems. Less-developed countries will likely argue that it is unfair for them to reduce their emissions by the same percentage as the more-developed nations, or may even believe that they should be permitted to increase their emissions while the richer countries reduce theirs. An alternative approach would establish per capita limits that would require some countries to make major reductions while others would be required to make much smaller reductions, or perhaps even increase emissions. It would be an enormous task to get each country to agree on reductions that it considered to be fair, given differences in economic resources, sources of energy, and past pollution-reduction efforts.

A third approach would impose a carbon tax on all fossil fuel production. The tax would be set at the level that would result in whatever percent global reduction was required. One study has estimated, for example, that a tax of $113 per ton of carbon content, phased in over ten years, would result in a reduction of from 10 to 20 percent from current levels of CO_2. Estimates for alternative means of reducing greenhouse gas emissions give some idea of the kinds of costs that might be incurred in reducing the likelihood of global warming. One estimate calculated that the cost of imposing a $25 per ton energy tax on coal, designed to stabilize greenhouse gases, would increase power costs for industrial users by as much as 20 percent and for residential users, 15 percent. A $6 per barrel tax on oil would raise the price of gasoline at the pump by 17 to 20 cents a gallon and home heating oil by 20 to 25 cents per gallon. Annual fuel costs would increase by $85 to $100 and raise residential heating oil prices 28 to 36 percent.[71] Another study calculated that an equivalent tax of $5 per metric ton of CO_2 would raise coal prices by 10 percent, oil by 2.8 percent, and gasoline by 1.2 percent, and would result in a 13 percent reduction of greenhouse gases. A $100 per ton tax would raise coal prices by 205 percent, oil prices by 55 percent, and gasoline prices by 23.3 percent, for an expected decrease of 45 percent in greenhouse gas emissions.[72] These short-term costs might decrease as higher prices encourage conservation.

ALTERNATIVES FOR ACHIEVING INTERNATIONAL GOALS

A pollution tax, if it is high enough, can provide a strong incentive for companies to reduce their emissions in whatever way is most efficient for them—closing down some operations, using cleaner fuels, investing in control technologies, changing work practices, and so on. It also provides a clear inducement for reducing emissions below permitted levels. Under current regulatory approaches, companies gain no economic advantage in emitting less pollution than is legally permitted them. With a pollution tax, they would save money every time they reduced emissions. Pollution taxes operate like other cost factors in production, and are consistent with the kind of flexibility and autonomous decision making that are important to businesses. Agency officials would also have a motive for enforcing the law since the tax generates additional revenues for the agency. Companies that comply with the law would support strong enforcement efforts aimed at their competitors.[73]

A properly designed scheme of tradable permits for greenhouse gases, would increase overall economic efficiency and equalize the marginal costs of emissions reductions. It would permit the setting of emission limits while creating incentives to reduce emissions below limits, increase opportunities for sharing costs, and redistribute pollution rights that are being inefficiently used. However, tradable permits would require considerable investment in starting up monitoring and reporting systems that do not now currently exist in LDCs. There is little certainty about what levels of emissions are environmentally sustainable, how to allocate initial allowances, and how to reduce uncertainty about annual allowances.

In contrast, emissions taxes have a great number of important advantages. They are simpler to administer. They permit more flexible adjustment in response to new knowledge. They create a continuous incentive to reduce emissions. Levels can be established and maintained in ways that reduce uncertainty for tax-payers. They expand the options available for redirecting economic activity in ecologically sustainable directions. They promote equity through generating revenue that can be used to fund cleanup projects elsewhere. They can be phased in gradually, both over time and in magnitude, to minimize disruption. However, although the cost per

Economic Incentives and Greenhouse Gases

unit of pollution is the same, the marginal cost of reducing emissions may vary, and achieving the maximum reduction at the minimum cost is not assured. It is difficult to assess the necessary level of taxes to cause a shift in behavior significant enough to resolve environmental problems. The political barriers are considerable; opposition by regulated industries may be tremendous.[74] To create a tax on greenhouse gas emissions, for example, participating countries might need to accomplish the following:

Index all greenhouse gases according to their relative heat-trapping potential;

Index all energy sources that release greenhouse gases according to the CO_2-equivalent release for each unit of energy produced;

Determine the tax for each unit of greenhouse gas;

Adjust the tax rate based on such factors as a nation's current or past contribution to the accumulation of greenhouse gases, its percentage of total global emissions, or per capita emissions;

Determine a timetable for phasing in the tax; and

Decide whether to provide tax relief for low-income tax payers to offset the impact of the tax increase.[75]

Since the production of energy is the major greenhouse gases, improving energy efficiency is probably the best way to reduce greenhouse gas emissions. Technologies are already available, and, in some cases, are already in place to produce electricity more efficiently and to reduce its consumption. Policy options include tax incentives to encourage conservation-related investments, changes in government procurement, and elimination of counterproductive practices and subsidies. Such actions will have economic and environmental benefits, as they reduce costs and make products more competitive in international markets. Countries that lead these efforts will be able to market cleaner and more efficient processes throughout the world. Efforts at conservation in transportation, industrial processes, and household heating, lighting, and cooling are the most important steps nations can take to lessen the threat of global warming.[76]

One estimate of the costs of preventing global warming concluded that it would cost less than $40 billion (0.2 percent of global income) a year to stabilize carbon dioxide emissions, if the costs were no more than $5 per ton of CO_2 equivalent. Substitutes for CFCs, for example, are believed to be technologically possible; their use would bring about a significant reduction in greenhouse gases for less than $5 per ton of carbon dioxide equivalent. Reducing CO_2 emissions from energy production by 10 percent is projected to cost less than $10/ton. The cheapest way to reduce carbon dioxide levels is to stop deforestation efforts that are themselves not very profitable. Taking the most cost-effective steps to reduce greenhouse gases (primarily reducing CFC production and uneconomic deforestation) would cost about $4 per ton or $6 billion a year, and would reduce greenhouse gas emissions by about 16 percent. However, as the amount of CO_2 reduced increases, the costs per ton increase dramatically, as substitutes become more expensive or are not available. The marginal cost of a 50 percent reduction in CO_2 emissions, assuming the use of current technologies, might be about $130 per ton. That is, if a tax of $130 per ton of CO_2 were imposed over time and current technologies were employed, emissions might decrease by one-half, but at a total cost of about $180 billion a year, or one percent of the world's economic output. If reductions were pursued immediately, they would be even more expensive.[77]

These calculations become more complicated if an attempt is made to discount the costs and benefits of future climate change in determining what actions should be taken now. Given the uncertainty associated with climate change, the most economically efficient step might be to make investments with high returns and earmark the proceeds for future preventive and adaptive efforts as they become more necessary. Those who believe that the scientific uncertainties surrounding global warming may soon be resolved argue that if preventive steps become necessary, they will be more economically efficient if the problems are better understood.[78] Such a view, of course, assumes that warming trends can be reversed, and largely ignores concerns of intergenerational distribution of benefits and burdens. It is difficult to estimate how high taxes would have to be to achieve the required reduction, since energy use is a function of so many factors. Tax revenues could be used to help finance new, less polluting technologies in the poorer nations.

Another possible approach would establish a system of tradable emission permits that would allocate to each country the number of permits equal to the number of tons it could emit each year. The number of permits might initially be the same as current emission levels, and then ratcheted down on a regular basis until the desired reduction is achieved. Countries could buy and sell permits, so that overall reductions would be achieved at the lowest cost.[79] Countries might be more favorably disposed to such a scheme than to a carbon tax, but distributional questions remain. Would the wealthy nations buy up the emission permits and preclude economic growth in the developing world?

Another option is to redirect the annual $90 billion in foreign direct investment by private sources. World Bank studies have encouraged investments that go beyond traditional areas of economic and industrial output to build human capital through education, and social capital through strengthening local institutions and the capacity for collective problem solving.[80] The World Bank's 1996 Conference on Environmentally Sustainable Development focused on how to influence private lending in ways that contribute to environmentally sustainable development. However, few specific recommendations surfaced beyond simply appealing to investors to look at the long-term implications of their actions and to encourage them to invest in education, infrastructure, basic human services, and the institutions of civil society. The role of the state is not so much to provide assistance directly as to coordinate efforts of local institutions, provide technical and financial assistance, resolve conflicts, and build local capacity.[81]

If global agreements impose overall emission reduction goals on different countries, and leave it up to them to achieve those goals any way they choose, the agreements will raise important issues concerning the consequences of these policy choices for the competitive position of goods produced in participating countries. If a carbon tax was universally imposed, it would have minimal impact on the competitive position of products made in countries with similar energy production patterns. But this approach would nevertheless bring problems, as some countries have much greater fossil fuel resources than others, and their export markets would be adversely affected. Countries such as France that rely heavily on nuclear power would benefit, since products produced there would not be nearly so affected by the price increase of a carbon tax. A broader tax on energy could overcome some of these concerns.

The Importance of Decentralized Approaches to Implementation

Changes in regulatory instruments will do little to alter the overall context in which environmental regulation takes place, but they can contribute to more democratic decision making and greater integration of economic and environmental considerations. Regulatory approaches that emphasize dissemination of information and enhanced consumer choices, for example, can lead to increased awareness and understanding of our environmental options and engage more of us in explicitly making those decisions. Disclosure of information about pollution is one of the most powerful means available to reduce pollution. Some instruments, such as emission fees and taxes, can add the cost of environmental damage to the total cost of production so that market decisions are more likely to reflect true costs. In a political economy fundamentally committed to market exchanges, regulatory strategies that help ensure that the real costs of production are included in the prices charged can make a critical contribution to achieving environmental protection goals.[82]

Solving global environmental problems ultimately requires local efforts. Some argue that we need new ways of preserving the balance and harmony of society and nature. Small, intimate, local, autonomous, self-governing communities are more likely to learn to live in this way than are large states with the opposite characteristics. National and international standards can provide broad guidelines for action and create economic incentives, but effective efforts will need to come from local, decentralized ones.[83] Wendell Berry has warned us that "the real work of planet-saving will be small, humble, and humbling. . . . Its jobs will be too many to count, too many to report, too many to be publicly noticed or rewarded, too small to make anyone rich or famous."[84] The roots of our environmental problems are not political, but cultural. "The question that must be addressed," Berry has written,

> is not how to care for the planet, but how to care for each of the planet's millions of human and natural neighborhoods, each of its millions of small pieces and parcels of land, each one of which is in some precious way different from all the others. Our understandable wish to preserve the planet must somehow be reduced to the scale of our competence—that is, to the wish to preserve all of its humble households and neighborhoods.[85]

Economic approaches that promise flexibility and decentralization are important elements of effective implementation, but they are not enough to engender the kind of changes that may be required to preserve the global environment. A primary challenges of implementation is to ensure that economic incentives and other incremental steps move us in the direction of more fundamental change.

References

1. General Accounting Office, "Environmental Protection Agency" Transition Series (December 1992): 6–7.
2. Ibid., 10–11.
3. National Performance Review, *Creating a Government that Words Better and Costs Less: Reinventing Environmental Management* (Washington, D.C.: U.S. Government Printing Office, September 1993), 1–2.
4. Paul Portney, ed., *Public Policies for Environmental Protection* (Washington, D.C.: Resources for the Future, 1990).
5. Clifford S. Russell, "Monitoring and Enforcement," in *Public Policies for Environmental Protection*, ed. Paul Portney (Washington, D.C.: Resources for the Future), 243–74.

6. Ibid., 248–53.
7. Ibid., 263–64.
8. Ibid., 268–70.
9. George W. Bilicii, Jr., "Note, An Analysis of the Land Disposal Ban in the 1984 Amendments to the Resource Conservation and Recovery Act," *Georgetown Law Journal* 67 (1988): 1563.
10. Marcia E. Williams and Jonathan Z. Cannon, "Rethinking the Resource Conservation and Recovery Act for the 1990s," *Environmental Law Reporter* 21 (1991): 10063–68.
11. General Accounting Office, "Hazardous Waste: New Approach Needed to Manage the Resource Conservation and Recovery Act" (Washington, D.C.: General Accounting Office, 1988). See also Environmental Protection Agency, "The Nation's Hazardous Waste Management Program at a Crossroads: The RCRA Implementation Study" (1990).
12. See generally Cass Sunstein, *After the Rights Revolution* (Cambridge: Harvard University Press, 1990), 74–110; Stephen Breyer, *Regulation and its Reform* (Cambridge, Mass.: Harvard University Press, 1982); Bruce Ackerman and William Hassler, *Clean Coal/Dirty Air* (New Haven, Conn.: Yale University Press, 1981); Bruce Yandle, *The Political Limits of Environmental Regulation* (New York: Quorum, 1989); and Thomas Schelling, *Incentives for Environmental Protection* (Cambridge, Mass.: MIT Press, 1983).
13. Allen V. Kneese and Charles L. Schultze, *Pollution, Prices, and Public Policy* (Washington, D.C.: Brookings Institution, 1975).
14. Ibid., 16.
15. For a review of the history and evolution of environmental economics, a discussion of how to use taxes in determining optimal levels of pollution, and a primer on the measurement of environmental damage, see David W. Pearch and R. Kerry Turner, *Economics of Natural Resources and the Environment* (Baltimore: Johns Hopkins University Press, 1990).
16. Charles L. Schultze, *The Public Use of Private Interest* (Washington, D.C.: Brookings Institution, 1977), 18.
17. Thomas Schelling, *Incentives for Environmental Protection* (Cambridge, Mass.: MIT Press, 1983).
18. Bruce A. Ackerman and Richard B. Stewart, "Comment: Reforming Environmental Law," *Stanford Law Review* (1985): 1334–65.
19. Bruce Ackerman and Richard Stewart, "Reforming Environmental Law: The Democratic Case for Market Incentives," *Columbia Journal of Environmental Law*, 13 (1988): 171. An even more ambitious use of such marketlike mechanisms as emissions trading in environmental

References

regulation has been proposed by Anderson and Leal, who argue that markets should be created to determine the number of emission credits to be bought and sold. Persons who pollute and those who are affected by the pollution can come together and bargain for acceptable pollution levels. See Terry L. Anderson and Donald R. Leal, *Free Market Environmentalism* (San Francisco: Pacific Research Institute for Public Policy, 1991), 22.

20. See, generally, Government of Canada, "Economic Instruments for Environmental Protection: Discussion Paper" (1992), 9.
21. Ibid., 8.
22. See, generally, National Commission on the Environment, *Choosing a Sustainable Future* (Washington, D.C.: Island Press, 1993), 21–44.
23. Many countries have had some experience with emissions fees for pollutants released into lakes and streams. Germany, for example, has made wider use of effluent taxes and fees than has the U.S., particularly in the area of water pollution. *International Environmental Reporter* (2 February 1986): 53–55.
24. See, generally, Richard B. Stewart, "Controlling Environmental Risks through Economic Incentives," *Columbia Journal of Environmental Law* 14 (1988): 153–69.
25. Ibid.
26. Herman E. Daly, "Farewell Lecture to World Bank" (14 January 1994), 6.
27. Robert W. Hahn, "Innovative Approaches for Revising the Clean Air Act," *Natural Resources Journal* 28 (Winter 1988): 174.
28. For a discussion of tradable permits, see Robert W. Hahn and Gordon L. Hester, "Marketable Permits: Lessons for Theory and Practice," *Ecology Law Quarterly* 16 (1989): 36; Robert W. Hahn and Gordon L. Hester, "Where Did All the Markets Go? An Analysis of EPA's Emissions Trading Program," *Yale Journal on Regulation* 6 (1989): 109–153; Daniel J. Dudek and John Palmisano, "Emissions Trading: Why Is This Thoroughbred Hobbled?" *Columbia Journal of Environmental Law* 13 (1988): 217–256; and Clifford S. Russell, "Controlled Trading of Pollution Permits," *Environmental Science and Technology* 15(1 January 1981): 24–28.
29. See, generally, Environmental Protection Agency, *The United States Experience with Economic Incentives to Control Environmental Pollution* (Washington, D.C., 1992); Senators Timothy E. Wirth and John Heinz, *Project 88-Round II, Incentives for Action: Designing Market-Based Environmental Strategies* (Washington, D.C.); Environmental Protection Agency, "The New Generation of Environmental Protection," working draft (15 April 1994).

30. In *Chevron, U.S.A., Inc. v. Natural Resources Defense Council* the Supreme Court upheld the EPA regulations providing for a plantwide definition of stationary sources as a permissible, reasonable policy decision within the discretion of the agency. 467 U.S. 837 (1984). In 1986 the EPA issued an "Emissions Trading Policy Statement" that outlined requirements states must meet in creating, using, and banking emission reduction credits. 51 *Fed. Reg.* 43,814 (1986).
31. For discussion of proposals for market-based instruments that have resulted from the Clean Air Act, see Environmental Defense Fund and General Motors, "Mobile Emissions Reduction Crediting" (n.d.); Environmental Protection Agency, "Market-Based Incentives and Other Innovations For Air Pollution Control" (Research Triangle Park, N.C., June 1992). See also Gary C. Bryner, *Blue Skies, Green Politics: The Clean Air Act of 1990 and Its Implementation* (Washington, D.C.: Congressional Quarterly Press, 1995).
32. See, for example, Georgia Institute of Technology, "Emissions Banking and Trading: A Survey of U.S. Programs" (April 1994).
33. Illinois EPA, "Feasibility Study for Market-Based Approaches to Clean Air" (September 1993); Illinois EPA, "Draft Proposal: Design for NO_x Trading System" (22 September 1993).
34. "Illinois Begins Pilot Project to Scrap Old, Polluting Cars in Chicago," *Clean Air Report* (8 October 1992): T-9.
35. Palmer Bellevue Corporation, "Emissions Reduction Trading in the Chicago Metropolitan Area: A Pre-Feasibility Analysis of the Effectiveness of Market Measures to Control Ground-Level Ozone" (Chicago, May 1992).
36. Peter Jensen, "'Gas Guzzler' Law's Value as Tax Source Questioned," *Baltimore Sun*, 22 April 1992, 1A.
37. "Northeast Proposes to Cut Compliance Costs, Shield Region's Economy"; "Businesses Offer Plan For Banking, Trading to Cut Costs, Allow Growth," *Clean Air Report* (8 October 1992): T-10, T-9.
38. Texas Air Control Board, "Marketable Permits Feasibility Study" (June 1993).
39. South Coast Air Quality Management District, "Regional Clean Air Incentives Market, Summary Recommendations" (Spring 1992).
40. Richard W. Stevenson, "Trying a Market Approach to Smog," *New York Times*, 25 March 1992, C1.
41. South Coast Air Quality Management District, "Regional Clean Air Incentives Market, Established the Foundation" (Spring 1992), 2-2-2-3, 3-4.
42. B. Drummond Ayres, Jr., "California Smog Cloud Reveals a Silver Lining," *New York Times*, 3 November 1995, A7. However, the EPA

References

proposed new ozone and particulate air quality standards in 1997, complicating the efforts to comply with national standards.
43. Clifford S. Russell, "Economic Incentives in the Management of Hazardous Wastes," *Columbia Journal of Environmental Law* 13 (1988): 257, 265–70.
44. Ibid., 270–71.
45. Ibid., 272–73.
46. Joseph G. Morone and Edward J. Woodhouse, *Averting Catastrophe: Strategies for Regulating Risky Technologies* (Berkeley: University of California Press, 1986).
47. Ted Wett, "In the Beginning: Minimization Is Targeted as the Top Priority in an Effective Waste Management Program," *Chemical Marketing Reporter* 240 (18 November 1991): SR16.
48. Ibid.
49. William F. Pedersen, Jr., "The Limits of Market-Based Approaches to Environmental Protection," *Environmental Law Review* (April 1994): 10173–76.
50. For a discussion of tradable permits, see Hahn and Hester, "Marketable Permits: Lessons for Theory and Practice," and Hahn and Hester, "Where Did All the Markets Go?" 109–53.
51. Pederson, "The Limits of Market-Based Approaches," 141–42.
52. See Sloane E. Anders, "Note, The Federal Tax System and the Environment: Should Payments Made Pursuant to CERCLA Be Deductible?" *Virginia Tax Review* 10 (1991): 707.
53. Daniel R. Mandelker and Roger A. Cunningham, *Planning and Control of Land Development* (Charlottesville, Va.: Michie, 1990), 544.
54. See, generally, Marlin K. Smith, "From Subdivision Improvement Requirements to Community Benefit Assessments and Linkage Payments: A Brief History of Land Development Exactions," *Law and Contemporary Problems* 50 (1987): 5.
55. Kathleen M. Martin, "Public/Private Cooperation in the Development of Contaminated Properties," *American Bar Association Section on Real Property, Problems and Trusts Law* (1990), Tab C, at 9–10.
56. 59 FR 16690 (7 April 1994), 16991.
57. Vaclav Smil, *Global Ecology: Environmental Change and Social Flexibility* (London: Routledge and Keyan Paul, 1993), 215.
58. Daniel A. Seligman, "Air Pollution Emissions Trading: Opportunity or Scam?" (Unpublished paper, Sierra Club, Washington, D.C., April 1994), 8–10.
59. Letter from Gail Ruderman Feuer and Veronica Kun, Natural Resources Defense Council, to California Air Resource Board (8 March 1994).

60. Bureau of National Affairs, "California: Problems Seen Setting Baseline Levels in South Coast Emissions-Trading Program," *Environment Reporter*, 29 May 1992, 437.
61. California Air Resources Board, rule 2015(c) (1) provides only that the Executive Officer must propose to the Governing Board that increased allocations be offset from non-RECLAIM sources identified in California's Air Quality Management Plan. See letter from Feuer and Kun to California Air Resources Board (8 March 1994), 3.
62. Seligman, "Air Pollution Emissions Trading," 30.
63. See Anderson and Leal, *Free Market Environmentalism*.
64. *Clean Air Act*, Sec. 110(a)(2), 172(a)(2), 181(a), 186(a), 188(c), and 192(a) of the, 42 U.S.C. 7401 et seq. See Natural Resource Defense Council, "Comments Before the U.S. E.P.A. on Economic Incentive Program Rules and Related Guidance" (13 June 1993).
65. Praveen K. Amar, Michael J. Bradley, and Donna M. Boysen, "A Market-Based NOx Emissions Cap System in the NESCAUM Region," *Regulatory Analyst* 1:3 (March 1993): 1–7.
66. See, generally, Barry Commoner, *Making Peace With the Planet* (New York: Pantheon, 1990), and Steven Kelman, *What Price Incentives?* (Boston: Auburn House, 1981).
67. Thomas Michael Power and Paul Raber, "The Price of Everything," *Sierra* (November/December 1993): 87–96.
68. Michael Gartner, "A Skeptic Speaks," *EPA Journal* 18:2 (May/June 1992): 26.
69. Power and Raber, "The Price of Everything," 92–95.
70. Susan Okie, "Developing World's Role in Global Warming Grows," *Washington Post*, 15 May 1990.
71. Center for Strategic and International Studies, "Implications of Global Climate Policies" (Washington, D.C.: C.S.I.S., 1989), 8–11.
72. William D. Nordhaus, "Global Warming: Slowing the Greenhouse Express," in *Setting National Priorities: Policy for the Nineties*, ed. Henry J. Aaron (Washington, D.C.: Brookings Institution, 1990), 203–04.
73. Clifford S. Russell, "What Can We Get from Effluent Charges?" (Washington, D.C.: Resources for the Future, 1977).
74. Francisco R. Sagasti and Michael E. Colby, "Eco-Development and Perspectives on Global Change from Developing Countries," in *Global Accord: Environmental Challenges and International Responses*, ed. Nazli Choucri (Cambridge, Mass.: MIT Press, 1993), 175–203, at 198.
75. Ibid., 197.
76. World Resources Institute, *World Resources, 1990–91* (Washington, D.C.: World Resources Institute, 1990), 25–6.

References

77. Nordhaus, "Global Warming," 197–99.
78. Ibid., 205–07.
79. Michael Grubb, *The Greenhouse Effect: Negotiating Targets* (London: Royal Institute of International Affairs, 1989).
80. The World Bank, *Monitoring Environmental Progress: A Report on Work in Progress* (Washington, D.C.: World Bank, 1995), 57–64.
81. The World Bank's conference, "Rural Well-Being: From Vision to Action," was held in September 1996; proceedings of the conference will likely be available from the Bank by the end of 1997.
82. For a discussion of how economic analysis can be used to devise more efficient policies to reduce pollution, see Tom Tietenberg, *Environmental and Natural Resource Economics*, 3d. ed. (New York: Harper Collins, 1992), 360–391; Frances Cairncross, *Costing the Earth: The Challenge for Governments, The Opportunities for Business* (Boston: Harvard Business School Press, 1993), 89–110; and Anderson and Leal, *Free Market Environmentalism*.

    For a review of the experience of nations with market-based approaches to regulation, see Sanford E. Gaines and Richard A. Westin, eds., *Taxation for Environmental Protection: A Multinational Legal Study* (New York: Quorum Books, 1991); Glenn Jenkins and Ranjit Lamech, *Green Taxes and Incentives Policies* (San Francisco: ICS Press, 1994); and Organisation for Economic Co-operation and Development, *Environment and Taxation: The Cases of the Netherlands, Sweden and the United States* (Paris: OECD, 1994).

    For examples of how market-based instruments can be applied to global environmental problems, see Committee for Economic Development, *What Price Clean Air? A Market Approach to Energy and Environmental Policy* (New York: Committee for Economic Development, 1993), 46–82; Ernst U. Von Weizsacher and Jochen Jesinghaus, *Ecological Tax Reform* (London: Zed Books, 1992); Robert Repetto, Roger C. Dower and Robin Jenkins, and Jacqueline Geoghegan, *Green Fees: How a Tax Shift Can Work for the Environment and the Economy* (Washington, D.C.: World Resources Institute, 1992); Roger C. Dower and Mary Beth Zimmerman, *The Right Climate For Carbon Taxes: Creating Economic Incentives to Protect the Environment* (Washington, D.C.: World Resources Institute, 1992); Organisation for Economic Co-operation and Development, *Taxation and the Environment* (Paris: OECD, 1993); Organisation for Economic Co-operation and Development, *Taxing Energy: Why and How* (Paris: OECD, 1993); and Organisation for Economic Co-operation and Development, *Environmental Policy: How to Apply Economic Instruments* (Paris: OECD, 1991).

83. William Ophuls and A. Stephan Boyan, Jr., *Equality and the Politics of Scarcity Revisited* (New York: Freeman, 1992).
84. Wendell Berry, "Out of Your Car, Off Your Horse," *The Atlantic* (February 1991): 62.
85. Berry, Wendell, *What Are People For?* (San Francisco: North Point, 1990), 200.

4

International Institutions for Protecting the Environment

FIFTY YEARS AGO, during the closing days of the Second World War, the United Nations was created. The name for the new organization was taken from a statement by Franklin Delano Roosevelt and was first used in a 1942 declaration in which twenty-one nations pledged to work together to fight the Axis Powers. The United Nations Charter was drafted by representatives from China, France, the Soviet Union, Great Britain, and the United States who began meeting in Washington, D.C. in August 1944. They finished their work on June 26, 1945, when the charter was signed by delegates from fifty nations. Membership is open to all "peace-loving" nations.[1]

The UN General Assembly, composed of representatives from each member state, meets in regular session from the third Tuesday in September until mid-December. The session begins with the election of a president, twenty-one vice presidents, and chairs of the seven main working committees: disarmament and related international security issues; special political matters; economic and financial matters; social, humanitarian, and cultural matters; decolonization; administration and budget; and legal matters. Most of the 150 or so questions taken up by the General Assembly are divided among the committees.²

The UN Charter reflects an ambitious agenda:

WE THE PEOPLES OF THE UNITED NATIONS, DETERMINED

to save succeeding generations from the scourge of war, which twice in our lifetime has brought untold sorrow to mankind, and

to reaffirm faith in fundamental human rights, in the dignity and worth of the human person, in the equal rights of men and women and of nations large and small, and

to establish conditions under which justice and respect for the obligations arising from treaties and other sources of international law can be maintained, and to promote social progress and better standards of life in larger freedom,

AND FOR THESE ENDS

to practice tolerance and live together in peace with one another as good neighbours, and

to unite our strength to maintain international peace and security, and

to ensure by the acceptance of principles and the institution of methods, that armed force shall not be used, save in the common interest, and

to employ international machinery for the promotion of the economic and social advancement of all peoples

HAVE RESOLVED TO COMBINE OUR EFFORTS TO ACCOMPLISH THESE AIMS.

One measure of the UN's agenda is its special observances:

UN Decade of Disabled Persons	1983–1992
Second Decade to Combat Racism and Racial Discrimination	1983–1993

Transport and Communications Decade for Asia and the Pacific	1985–1994
World Decade for Cultural Development	1988–1997
International Decade for Natural Disaster Reduction	1990s
Second Industrial Development Decade for Africa	1990s
Third Disarmament Decade	1990s
United Nations Decade of International Law	1990–1999
International Decade for the Eradication of Colonialism	1991–2000
Second Transport and Communications Decade in Africa	1991–2000
United Nations Decade Against Drug Abuse	1991–2000
International Space Year	1992
International Year for the World's Indigenous Peoples	1992
International Year of the Family	1994
Fourth United Nations Development Decade	1991–2000[3]

These special efforts are in addition to the United Nation's primary functions of peace keeping and humanitarian relief efforts.

The United Nations has been widely criticized as obsolete and ineffective. It has been a popular target of criticism by isolationists in the United States who fear loss of sovereignty or other threats arising from international government or the "new world order." For others, the UN continues to represent the great promise of global peace and serves as the vehicle for caring for the earth, ending hunger and poverty, promoting justice and fairness, and increasing opportunities and hope for the children of the world.[4]

Many observers and policy makers call for strong, centralized agencies to implement global agreements, while others call for a much more decentralized approach. Some argue that these tasks should be added to the jurisdictions and agendas of current international and national bodies, while others argue for the creation of new bodies. What kinds of international institutions are needed to ensure effective implementation of global agreements? Are the United Nations and other existing institutions capable of fulfilling these roles, or are new institutions needed?

An Overview of International Institutions

Several international organizations play at least some role in international environmental policy. Environmental protection efforts are intertwined with other efforts, particularly international development, since many international organizations provide assistance to the less developed nations. As discussed in Chapter 6, international development and environment protection efforts are closely related: Development projects may exacerbate environmental problems or contribute to their resolution, and environmental improvements are often prerequisites to improving health, agricultural output, and economic activity. As a result, both development and environmental organizations are part of the institutional infrastructure for dealing with global environmental problems. One study has identified more than 120 such agencies and programs.[5] Table 4.1 lists a few of the organizations and programs that deal with global environmental issues.

A detailed discussion of each of these organizations is beyond the scope of this chapter. Given the importance of the United Nations and affiliated agencies in formulating and implementing international environmental agreements, the discussion focuses on that group. The problems of overlapping jurisdictions and competing efforts within the United Nations system are considerable, and they are greatly magnified in light of the whole range of institutions and efforts for global environmental protection.

An Overview of International Institutions

TABLE 4.1
SELECTED INTERNATIONAL ORGANIZATIONS FOR THE ENVIRONMENT

United Nations
  Centre for Human Settlements
  Conference on Trade and Development
  Development Programme
  Environment Programme (Global Environmental Monitoring System)
  Fund for Population Activities
  Habitat and Human Settlements Foundation
  Industrial Development Organization
  Institute for Training and Research
  Scientific Committee on Effects of Radiation
  International Court of Justice (World Court)
United Nations–Affiliated Organizations
  Food and Agricultural Organization
  Intergovernmental Oceanographic Commission
  International Atomic Energy Agency
  United Nations Educational, Scientific, and Cultural Organization
    Man and the Biosphere Program
  World Health Organization
    International Agency for Research on Cancer
  World Meteorological Organization
Non-UN Governmental
  European Community
  International Council for the Exploration of the Sea
  International Energy Agency
  North Atlantic Treaty Organization Committee on Challenges of Modern Societies
  Organisation for Economic Co-operation and Development
  South Asia Cooperative Environment Program
Nongovernmental Scientific, Technical, and Professional
  Environmental Law Center
  Environmental Liaison Center (Nairobi)
  European Environmental Bureau (Brussels)
  International Association for Ecology
  International Council of Scientific Unions
  International Social Science Council

(Continued on page 178)

INSTITUTIONS FOR PROTECTING THE ENVIRONMENT

(TABLE 4.1 *Continued from page 177*)

SELECTED INTERNATIONAL ORGANIZATIONS FOR THE ENVIRONMENT

International Union for Conservation of Nature and Natural Resources
  International Council for Environmental Law
  International Institute for Environment and Development
  International Ocean Institute
  International Nuclear Information System
  World Environment Center (New York)
  World Wildlife Fund
International Funding Agencies
  African Development Bank
  Inter-American Development Bank
  International Bank for Reconstruction and Development (World Bank)

SOURCE: Adapted from Lynton Keith Caldwell, *International Environmental Policy* (Durham, N.C.: Duke University Press, 1984), 282–86.

The UN System

The United Nations General Assembly includes representatives from every member state. It is empowered to adopt resolutions, that, while not binding on members, carry some weight as a statement of world opinion. The Security Council, composed of five permanent and ten two-year rotating-term members, can adopt, with at least nine votes (including those of all five permanent members), resolutions that are binding on all members. The Secretariat, headed by the Secretary General, is the administrative body that carries out the business of the organization and performs other ministerial acts. The Economic and Social Council was created under the UN Charter as the principal body to coordinate economic and social activities, and it oversees a bewildering array of agencies and organizations, including the following:[6]

UNRWA United Nations Relief Works and Agency for PalestineRefugees in the Near East
IAEA International Atomic Energy Agency
INSTRAW International Research and Training Institute for the Advancement of Women

An Overview of International Institutions

UNCHS	United Nations Centre for Human Settlements (Habitat)
UNCTAD	United Nations Conference on Trade and Development
UNDP	United Nations Development Program
UNEP	United Nations Environment Program
UNFPA	United Nations Population Fund
UNHCR	Office of the United Nations High Commissioner for Refugees
UNICEF	United Nations Children's Fund
UNITAR	United Nations Institute for Training and Research
UNU	United Nations University
WFC	World Food Council
WFP	Joint UN-FAO World Food Program

The Council supervises five regional commissions:

Economic Commission for Africa (ECA)

Economic Commission for Europe (ECE)

Economic Commission for Latin America and the Caribbean (ECLAC)

Economic and Social Commission for Asia and the Pacific (ESCAP)

Economic and Social Commission for Western Asia (ESCWA)

It directs the activities of sessional, standing, and ad hoc committees, as well as six functional commissions:

Commission for Social Development

Commission on Human Rights

Commission on Narcotic Drugs

Commission on the Status of Women

Population Commission

Statistical Commission

Finally, the Council oversees nearly twenty specialized organizations with key environmental and social responsibilities:

ILO	International Labor Organization
FAO	Food and Agriculture Organization of the United Nations
UNESCO	United Nations Educational, Scientific and Cultural Organization
WHO	World Health Organization
IBRD	International Bank for Reconstruction and Development (World Bank)
IDA	International Development Association
IFC	International Finance Corporation
IMF	International Monetary Fund
ICAO	International Civil Aviation Organization
UPU	Universal Postal Union
ITU	International Telecommunication Union
WMO	World Meteorological Organization
IMO	International Maritime Organization
WIPO	World Intellectual Property Organization
IFAD	International Fund for Agricultural Development
UNIDO	United Nations Industrial Development Organization
GATT	General Agreement on Tariffs and Trade

Another way to capture the extraordinary complexity of the UN system is to list organizations by environmental areas of responsibility as shown in Box 4.1.

The sheer number of agencies means that the UN will continue to be at the center of any global environmental effort; any suggestion that the UN can be replaced by a new set of institutions seems unrealistic. The challenge is how to make some sense of this array of organizations and ensure that their efforts are at least somewhat coordinated and compatible. That is no easy task, as a brief review of some of the major institutions shows.

An Overview of International Institutions

BOX 4.1

UN BODIES WITH ENVIRONMENTAL RESPONSIBILITIES

Atmosphere

Organizations with Major Programs
World Meteorological Organization (WMO)
United Nations Environment Program (UNEP)
United Nations Industrial Development Organization (UNIDO)
International Maritime Organization (IMO)
United Nations Educational, Scientific and Cultural Organization (UNESCO)
 /Inter-governmental Oceanographic Commission (IOC)
United Nations Development Program (UNDP)
International Arctic Science Committee (IASC)

Advisory Bodies
Inter-governmental Panel on Climate Change

Agriculture/Land

Organizations with Major Programs
Food and Agriculture Organization (FAO)
United Nations Environment Program (UNEP)
United Nations Children's Fund (UNICEF)
World Food Program (WFP)
United Nations Development Program (UNDP)
International Labor Organization (ILO)
United Nations Sudano-Sahelian Office and UN Statistical Office (UNSO)
United Nations Educational, Scientific and Cultural Organization
   Man and the Biosphere Program (MAB)

Coordination Arrangements
Administrative Committee on Coordination Task Force on Rural Development
Interagency Working Group on Desertification

Advisory Bodies
Consultative Group on International Agricultural Research
International Board for Plant Genetic Resources

(*Continued on page 182*)

(BOX 4.1 *Continued from page 181*)

Forests

Organizations with Major Programs
Food and Agriculture Organization (FAO)
World Food Program (WFP)
United Nations Environment Program (UNEP)
United Nations Development Program (UNDP)
United Nations Educational, Scientific and Cultural Organization (UNESCO)

Coordination Arrangements
Tropical Forests Action Program (TFAP)

Biodiversity Biotechnology

Organizations with Major Programs
Food and Agriculture Organization (FAO)
United Nations Environment Program (UNEP)
United Nations Industrial Development Organization (UNIDO)
United Nations Development Program (UNDP)
United Nations Educational, Scientific and Cultural Organization (UNESCO)

Coordination Arrangements
Ecosystem Conservation Group (ECG)

Advisory Bodies
International Board for Plant Genetic Resources ((BPGR)
Consultative Group on International Agricultural Research (CGIAR)

Oceans and Coasts

Organizations with Major Programs
United Nations Educational, Scientific and Cultural Organization (UNESCO)
 Inter-governmental Oceanographic Commission (IOC)
Food and Agriculture Organization (FAO)
United Nations Environment Program (UNEP)
International Maritime Organization (IMO)
United Nations Conference on Trade and Development (UNCTAD)
United Nations Development Program (UNDP)
World Meteorological Organization (WMO)
International Labor Organization (ILO)
International Civil Aviation Organization (ICAO)
International Telecommunications Union (ITU)
United Nations Industrial Development Organization (UNIDO)
World Health Organization (WHO)

An Overview of International Institutions

Coordination Arrangements
Office for Ocean Affairs and the Law of the Sea (OALOS)
Inter-Secretariat Committee on Scientific Problems Relating to Oceanography (ICSPRO)

Advisory Bodies
Group of Experts on the Scientific Aspects of Marine Pollution (GESAMP)

Freshwater

Organizations with Major Programs
United Nations Educational, Scientific and Cultural Organization (UNESCO)
World Health Organization (WHO)
World Meteorological Organization (WMO)
United Nations Environment Program (UNEP)
Food and Agriculture Organization (FAO)
United Nations Development Program (UNDP)

Coordination Arrangements
Administrative Committee on Coordination
Inter-secretariat Group on Water Resources

Toxic/Hazardous/Solid Wastes

Organizations with Major Programs
United Nations Environment Program (UNEP)
Food and Agriculture Organization (FAO)
International Labor Organization (ILO)
United Nations Development Program (UNDP)
United Nations Industrial Development Organization (UNIDO)
International Atomic Energy Agency (IAEA)

Population

Organizations with Major Programs
United Nations Population Fund (UNFPA)
Population Commission

Living and Working Environment for the Poor

Organizations with Major Programs
United Nations Commission for Human Settlements (UNCHS)
United Nations Development Program (UNDP)
World Health Organization (WHO)

(Continued on page 184)

> (BOX 4.1 *Continued from page 183*)
> Food and Agriculture Organization (FAO)
> United Nations Educational, Scientific and Cultural Organizations (UNESCO)
>   Man and the Biosphere Program (MAB)
> International Fund for Agricultural Development (IFAD)
> World Food Program (WFP)
> International Labor Organization (ILO)
>
> **Energy and Transport**
> *Organizations with Major Programs*
> New and Renewable Resources of Energy (NRSE)
> United Nations Development Program (UNDP)
>
> **Human Health**
> *Organizations with Major Programs*
> World Health Organization (WHO)
> United Nations Development Program (UNDP)
> Food and Agriculture Organization (FAO)
> United Nations Education, Scientific and Cultural Organization (UNESCO)
> World Meteorological Organization (WMO)
> International Fund for Agricultural Development (IFAD)
> International Atomic Energy Agency (IAEA)
> United Nations Children's Fund (UNICEF)
>
> **Environmental and Development Information**
> *Organizations with Major Programs*
> United Nations Environment Program (UNEP)
> World Health Organization (WHO)
> Food and Agriculture Organization (FAO)
> World Meteorological Organization (WMO)
> United Nations Sudano-Sahelian Office and United Nations Statistical Office (UNSO)

THE UNITED NATIONS ENVIRONMENT PROGRAM

The United Nations Environment Program (UNEP) is a subsidiary of the United Nations and has no independent legal basis. Its governing council is made up of fifty-eight members who meet every two years. Contributions to UNEP are voluntary, and the program

An Overview of International Institutions

United Nations Development Program (UNDP)
United Nations Conference on Trade and Development (UNCTAD)
United Nations Population Fund (UNFPA)
International Labor Organization (ILO)
United Nations Children's Fund (UNICEF)

Coordination Arrangements
Earthwatch
Advisory Committee for Coordination of Information Systems (ACCIS)

Environmentally Sound Technology

Organizations with Major Programs
United Nations Center for Science and Technology for Development (UNCSTD)
United Nations Environmental Program (UNEP)
United Nations Center on Transnational Corporations (UNCTC)
United Nations Conference on Trade and Development (UNCTAD)
United Nations Industrial Development Organization (UNIDO)
United Nations Development Program (UNDP)
United Nations Educational, Scientific and Cultural Organization

Advisory Bodies
Advisory Committee on Science and Technology for Development (ACAST)

Trade and Environment Economics

Organizations with Major Programs
General Agreement on Tariffs and Trade (GATT)
United Nations Conference on Trade and Development (UNCTD)
United Nations Development Program (UNDP)

SOURCE: Lee Kimball, *Forging International Agreement* (Washington, D.C.: World Resources Institute, 1992), 41.

has had a chronic problem with insufficient funds. Nevertheless, it played a major role in the negotiations leading to the stratospheric ozone agreement, and has sponsored meetings to discuss protecting regional seas, environmental impact assessments, the export of hazardous substances, and the protection of biodiversity. Headquartered in Nairobi, the UNEP seeks particularly to address the environmental dilemmas confronting developing countries. UNEP joined with the World Meteorological Organization to form an Intergovernmental Panel on Climate Change. This panel, composed of three working groups, is studying the causes, effects, and response strategies for global climate change.[7]

THE WORLD BANK AND INTERNATIONAL MONETARY FUND

The 1944 Bretton Woods conference on the global financial system resulted in the creation of the World Bank and the International Monetary Fund to help finance the reconstruction of Europe, which was devastated by World War II, and to provide a mechanism for promoting international trade. They both come under the broad UN umbrella because they have established formal, legal relationships with UN agencies, but they are otherwise autonomous bodies. These institutions are dominated by the more-developed nations, which contribute funds and have votes in proportion to their contribution.

The World Bank is made up of 180 nations, each of which has a financial share in the institution. The G7 countries—Canada, France, Germany, Italy, Japan, the United Kingdom, and the United States—control 47 percent of the Bank's resources. The Bank is made up of four institutions: The International Bank for Reconstruction and Development is the main lending institution; the International Development Association loans money, interest free, to the world's poorest nations; the International Finance Corporation lends money directly to the private sector; and the Multilateral Investment Guarantee Agency provides risk insurance to private foreign investors. The International Monetary Fund, like the World Bank, was created in 1944. It has 181 members that contribute to the fund on the basis of their economic wealth. Members can draw on IMF funds to support their currency when it is weak, borrow money to meet short-term foreign payment requirements, and receive loans.[8]

The World Bank is the largest source of funds for development projects. Its projects have been severely criticized as ecological

An Overview of International Institutions

disasters causing pollution and deforestation, requiring export agriculture rather than domestic food production, and forcing resettlement of indigenous peoples.[9] The International Monetary Fund has imposed economic adjustment programs on less-developed countries as a condition for receiving international financial assistance. These required "adjustments" include decontrol of exchange rates and currency, deregulation of foreign direct investment in LDCs, and trade liberalization. Unsustainable harvesting of natural resources, increased pollution, and social disruption have also been consequences of IMF programs.[10]

World Bank president James D. Wolfensohn announced in February 1997 a reorganization plan to improve the Bank's effectiveness. An internal study had concluded that only two-thirds of Bank-funded projects were successful—loans repaid, continued economic benefit, and no adverse environmental or social impacts. The proposal aims to increase the success rate to 75 percent by focusing on countries that are relatively stable and have a good track record for using aid effectively, by moving away from construction of big facilities such as power plants and highways to projects in health and education, and by increasing partnerships between the Bank and private investors. This initiative is the fifth major attempt to reorganize and redirect the Bank in a decade.[11] Critics of the World Bank argue that the promised reforms have had little impact on moving projects towards environmental sustainability. A coalition of 160 NGOs formed a "50 Years is Enough" campaign to attack the World Bank and the IMF's environmental record and the two organizations' demands for structural adjustment programs in the developing world that shift resources away from social needs.[12]

COMMISSION ON SUSTAINABLE DEVELOPMENT

The 1992 UNCED summit called for the creation of a new agency to monitor progress in improving environmental quality and the quality of life in the less developed world. In 1993, the UN General Assembly authorized the Economic and Social Council to establish a Commission on Sustainable Development (CSD) as a body within the council. The CSD was created to encourage and monitor progress in implementing the policies suggested in Agenda 21 and to encourage the integration of environmental protection and development efforts. The commission has no legal authority; its major role is to explore issues relevant to sustainable development, build consensus on needed actions, exchange information, and facilitate cooperative efforts. The commission, formally established in 1993 at a meeting of the United Nations Economic and Social Council, reports to that council. The commission makes recommendations to the UN General Assembly; a special session has been scheduled for 1997 to assess the progress in accomplishing the goals outlined in Agenda 21. The commission has fifty-three seats—thirteen for Western Europe and North America, thirteen for Africa, eleven for Asia, ten for Latin America and the Caribbean, and six for Eastern Europe.[13]

OTHER INSTITUTIONS

The 1992 Earth Summit and related meetings spawned other institutions. The United Nations Administrative Committee on Coordination, chaired by the Secretary General, brings together the heads of the major UN agencies, including the financial institutions, to coordinate implementation of the measures suggested in Agenda 21. The Inter-Agency Committee on Sustainable Development was created in 1992 to make recommendations to the Committee on Coordination for ways to improve coordination. Agenda 21 also called for the creation of a high-level advisory body to advise the UN and other intergovernmental institutions on these issues, and a twenty-one-member panel was established in 1993.[14]

Agenda 21 urged the UN to give greater voice to NGOs in UN agency deliberations—to let them participate in formal meetings and working group sessions, submit reports and recommendations, and participate in formal country delegations. Agenda 21 also recommended bolstering the United Nations Environment Program so that it can play a greater role in global environmental policy making. The United Nations Development Program is focusing on ways to strengthen institutional capacity for sustainable development and has added a new unit, called Capacity 21, to assist less-developed countries in related efforts. The United Nations Conference on Trade and Development is expected to focus on the interaction of trade, development, and environmental protection. UN regional economic commissions, regional development banks, regional offices of major UN agencies, and other regional organizations are part of the UN structure for sustainable development and environmental protection.[15]

Regional Organizations

Regional organizations are playing an increasingly important role in environmental protection, especially in Europe. For example, some twenty-five regional bodies are involved in programs aimed at protecting the earth's oceans, along with fourteen UN agencies that have regional divisions for oceans. The Baltic Marine Environment Protection Commission (also known as the Helsinki Commission, or HELCOM), for example, was established in 1980. In 1988 its members agreed to cut ocean pollution in half by 1995. The Nordic Investment Bank, owned by the Scandinavian nations, supports projects to clean up the Nordic environment. In 1990, eight European nations and the EC agreed to install the best available technology for their major industries and to improve municipal sewage treatment plants. The World Bank, European Investment Bank, European Bank for Reconstruction and Development, and Nordic Investment Bank all supported these efforts. In 1990, the Regional Environmental Center for Central and Eastern Europe, initially funded by the United States, Canada, and the European Community, was created to serve as a clearinghouse for information on Western environmental technologies and management strategies that might be useful to other nations. The center also assists governments in establishing effective environmental institutions and helps resolve conflicts between environmental groups and local governments.[16]

INSTITUTIONS FOR PROTECTING THE ENVIRONMENT

Several organizations in Western Europe promote cooperative efforts in environmental protection. The European Community (EC), created in 1957, has twelve members. A regional economic integration organization, the EC has traditionally focused on trade issues but has increasingly addressed environmental issues that are implicated in trade. The EC Commission is made up of fourteen commissioners who represent various interests of the community; one commissioner is specifically charged with environmental protection. The commission has the power either to propose regulations that become binding in member states or to issue directives to member states to enact domestic legislation. Environmental actions usually take the form of directives, since they give members some flexibility in deciding how to implement them and they reflect the diversity in members' legal systems.

The European Council, composed of representatives from member states, has the authority to adopt regulations and directives. Some issues require consensus, while others are determined by weighted majority vote. The council, which has no authority to propose actions, relies on the commission's technical expertise for policy initiatives. The European Parliament is directly elected by the public in each member state. Although it has been fairly involved in environmental issues and has prodded the commission and the council to take action, it is basically an advisory body. The European Court of Justice interprets EC law: Members can sue each other for failing to implement a directive or enforce a regulation. The UN Economic Commission for Europe deals primarily with economic issues, but has been the major vehicle for dealing with East-West issues in Europe and has addressed the long-range transportation of air pollutants in Europe. The Organisation for Economic Co-operation and Development (OECD), created in 1960, is made up of twenty-four nations, including the market-oriented industrial nations of Western Europe and the United States, Canada, Japan, Australia, and New Zealand. OECD sponsors research on a variety of issues, including comparisons of environmental policies of member states. It issues binding decisions and nonbinding recommendations.

An Overview of International Institutions

Central Europe's efforts to restore and manage its environment are augmented by new partnerships with Western governments and institutions. The Bank for European Reconstruction and Development (BERD) was created in 1990 to help Central European nations protect the environment. The bank has forty-two charter members and an initial budget of $12 billion. The European Investment Bank has funded projects in Eastern Europe, such as improving domestic gas production in Poland. The World Bank has also loaned money to Poland and Hungary to identify and redress environmental problems, establish a decentralized system for managing environmental problems, reform energy prices, help chemical and energy industries pollute less, develop energy conservation projects, and develop new energy sources. Other international organizations funnel financial assistance from the industrialized nations to the less-developed countries for environment and natural resource projects. As of 1990, according to one estimate, some $728 million had been given for such efforts ($105 million from the United States, $123 million from the European Community, and about $500 million from other industrialized countries).[17]

Nongovernmental Organizations (NGOs)

A host of nongovernmental actors are part of the quiltwork of global environmental institutions. Multinational corporations and large banks make decisions with enormous consequences for environmental quality and resource preservation, as well as economic growth and development. International associations of scientists support research that triggers demands for global action and provides a basis for shaping policies. Environmental and other coalitions can focus considerable public pressure on environmental problems through publicity, supplying information to governmental bodies, and facilitating environmental activism in various nations. Media decisions concerning what issues to address and how to depict them significantly shape global policy formulation and implementation. A few bodies, such as the World Conservation Union (IUCN) and the Ecosystem Conservation Group (ECG), combine national governments, international agencies, and NGOs.[18]

Western environmental groups and foundations have become increasingly active in Eastern Europe. The U.S. National Academy of Sciences has organized workshops with Polish and Romanian scientists on energy efficiency and natural resources management. A number of other groups, including the Institute for European Environmental Policy, Friends of the Earth International, World Environment Center, Management Sciences for Health, Environmental Law Institute, Rockefeller Brothers Fund, World Conservation Union, Ecological Studies Institute, and the International Institute for Applied Systems Analysis, are involved in a variety of environmental protection projects. The International Council of Scientific Unions has established a number of programs and task forces to explore climate change, ocean circulation, global atmosphere, and satellite cloud climatology.[19]

One of the newest NGOs is the Earth Council, created in late 1992 by Maurice Strong, secretary general of the Earth Summit, and others. The council, located in Costa Rica, is working with other NGOs to assess nations' efforts of nations to achieve sustainable development goals. One of the oldest is the Committee on International Development Institutions on the Environment, which is made up of UN and non-UN agencies and other global and regional bodies. This NGO was established in 1980 to review UN agencies' efforts at promoting sustainable development.[20] The Stockholm Initiative on Global Security and Governance led to the creation of the Commission on Global Governance, another organization that is working to improve the functioning of international environmental organizations.[21]

Assessing International Environmental Organizations

It is difficult to gauge the effectiveness of international organizations in addressing global environmental concerns. Many argue that such efforts are too little, too late; others find them premature, too expensive, or inefficient. Much of the disagreement is rooted in

Assessing International Environmental Organizations

our limited knowledge of ecosystems and how they respond to the threats confronting them.

The large number of environment-oriented organizations and programs reflects the widespread recognition of the importance of international environmental issues. There is, however, little coordination among these sometimes competing organizations, nor is there an overarching body to ensure that environmental concerns are given priority at the highest levels of policy making. No environmental organizations exist comparable to those for arms talks and international trade, for example; and despite the success of the 1992 Earth Summit in drawing attention to environment and development issues, environmental concerns still lack a central place in international relations. There have been few successful attempts to integrate environmental protection and development; development activities are themselves often plagued by competing and conflicting institutions, overlapping efforts, lack of coordination, and failure to integrate the less-developed nations into the world economy.[22]

As the discussion of the Law of the Sea and other global environmental agreements in Chapter 7 indicated, global agreements have been difficult to formulate and implement. Although the United Nations has been quite successful in providing a forum for debate, discussion, and negotiations, for the most part it lacks the ability to enforce agreements made under its jurisdiction. Critics call it a "bloated, clumsy organization" that has not really changed since its creation a half-century ago. There has been little turnover in the staff, and positions are based on patronage rather than merit. Multinational corporations, nongovernmental organizations, and financial institutions that play a critical role in international affairs are not represented in UN deliberations. In relying on consensus, the UN is required to water down efforts to meet the "least common denominator."[23]

The UN has been more visible in the 1990s than perhaps at any time since its creation. Its efforts at conference organizing, treaty negotiations, and peacekeeping have fostered great expectations, but its shortcomings seem more and more apparent. UN forces were embroiled in some thirteen peacekeeping missions in the early

1990s—in the Middle East (Egypt-Israel border), India-Pakistan, Cyprus, Golan Heights, Lebanon, Iraq-Kuwait, Angola, the former Yugoslavia, El Salvador, Western Sahara, Cambodia, Somalia, and Mozambique.[24] The peacekeepers are widely dispersed, underfunded, and often ineffective. The Angolan peace plan developed under UN auspices has collapsed into violence. UN peacekeepers have failed to keep Bosnia from being ravaged by war. Somalia relief degenerated into an effort to blame the UN. In the early 1990s, UN agencies were trying to care for 14 million refugees and to provide relief to countless victims of famine and war.[25]

A report by Dick Thornburg, the former U.S. Attorney General and governor of Pennsylvania who served as the UN's top management official, concluded that "duplication is widespread, coordination is often minimal, bureaucratic battles aimed at monopolizing a particular subject are rife and organizational objectives are sometimes in conflict." Former Secretary General Boutros-Ghali's efforts in administrative reform were hindered by bureaucratic infighting.[26] Demands for reforming the UN have become increasingly strong in the United States. They led to the demand by the Clinton administration in 1996 that Boutros Boutros-Ghali be ousted as UN Secretary General. Some four hundred projects were underway at the UN by the end of 1996 that were aimed at cutting costs and bringing expenditures within its $2.6 billion budget.[27]

The credibility of global agencies will be critical in gaining the largely voluntary support and participation of sovereign states. But the UN is viewed by some in the South as a tool of the wealthy nations; although the General Assembly may provide a forum for speakers from the LDCs, the MDCs control the funding and institutions of real power. The UN agenda is daunting: It spent $3.6 billion in 1992 on peacekeeping efforts and is strapped for funds. Only a small fraction of its 183 members have paid their dues in full. At the end of May 1995, UN members owed $2.75 billion in assessments, including $1.179 billion owed by the United States (the U.S. committed itself in 1972 to paying 25 percent of the UN's budget).[28] Dissatisfaction with the United Nations is growing in the

United States. One longtime observer of the UN warned that it "must be rebuilt as an organization capable of fulfilling its original purposes and dealing with the dangers that have grown up during its nearly half-century of existence—not just wars, but the economic stagnation and poverty that underlie so much of the world's conflicts." The former U.S. Ambassador to the UN, Madeleine K. Albright, addressed the General Assembly in 1993 by observing that "forty years of neglect have left this institution flabby and out of shape."[29]

A study of several regional environmental institutions concluded that "none of the institutions in the study was successful from the outset, though some eventually became so. In fact, most international environmental institutions were initially considered to be disappointing by their creators."[30] Two factors were particularly important in determining success. First, institution building is a dynamic process that begins with commitments to international norms or principles: "If states waited to form institutions until there was enough concern and scientific understanding to adopt strong rules, they would wait much too long. Institutions are needed early on to help create the conditions that make strong rules possible." Once this groundwork has been laid, institutions can respond to crises or other opportunities for more aggressive action and can formulate specific rules.[31] Second, the development of nonpartisan, widely accepted and acknowledged scientific assessments is critical. Given the dynamic nature of knowledge, institutions "should not codify existing knowledge in rules that are difficult to change, but should, on the contrary, foster an open-ended process of knowledge creation."[32]

The 1994–1995 World Resources Institute report concluded that "even with threats to the environment mounting, the UN system's shortcomings in terms of managing, coordinating, and evaluating international environmental and development issues have become more serious." The report focused on three concerns: (1) various UN agencies compete for scarce financial resources; (2) financial and policy commitments made by various agencies are uncoordinated;

INSTITUTIONS FOR PROTECTING THE ENVIRONMENT

and (3) despite the UN's original charter, the most important economic and financial decisions are made by other international bodies and at economic summits organized by the wealthy nations.[33]

Building Institutional Capacity for Environmental Protection

Agreements in other policy areas hold some lessons for the prospects of formulating and implementing global environmental agreements. Recent agreements concerning chemical weapons and nuclear arms reductions demonstrate some possibilities for the development of widely shared values that translate into prohibitions against or limitations on certain behaviors. Other agreements in trade and tariffs and international monetary exchanges illustrate the importance of domestic political and economic considerations in fashioning international agreements.

Because of growing fears of environmental calamities, many have called for a dramatic expansion of global institutional authority. The Stockholm Initiative on Global Security and Governance, for example, proposed in 1991 that the following steps be taken in global government to address environmental and related concerns:

> that the United Nations take on a broadened mandate at the Security Council level, following the wider understanding of security which has developed, and that its composition and the use of the veto be reviewed;
>
> that the Secretary-General be given a stronger position and the means to exercise authority, and that the method of appointment of the Secretary-General and of higher-level staff be reviewed;
>
> that the system-wide responsibilities and authority of the Secretary-General concerning interagency coordination and cooperation should be firmly established;
>
> that the financing system of the United Nations be reviewed, and that countries who do not adhere to the financial rules be deprived of the right to vote;
>
> that the activities of the United Nations in the economic and social fields be strengthened and rationalized;

Building Institutional Capacity for Environmental Protection

that the International Monetary Fund and the World Bank be coordinated, among themselves and with the United Nations system and GATT, with the aim of a clearer division of labour, better harmony and full universality in their work;

that a World Summit on Global Governance be called, similar to the meetings in San Francisco and at Bretton Woods in the 1940s; and

as a matter of priority, the establishment of an independent International Commission on Global Governance.[38]

The UNCED report reaffirmed many of these recommendations, suggesting that global agreements:

have more precisely stated objectives;

avoid excessive use of reservation clauses which affect the level of participation;

contain provisions that encourage speedy entry-into-force to allow for implementation;

establish more flexible procedures for settling disputes;

require the circulation of information regarding implementation;

include more expeditious procedures for amendments;

contain provisions that encourage, through additional protocols, further development of the law in the subject matter covered by the instrument;

provide for mechanisms for monitoring compliance and for periodic review of the treaty regime, relying on, where appropriate, streamlined, harmonized, or combined administrative or supporting services and institutions; and

improve the negotiating process of the instruments.

Other recommendations included specific institutional changes, particularly reform of the United Nations institutions in the field of environment and development:

proliferation of institutions at the global level must be avoided;

existing international institutions at the global and regional levels in the field of environment and development, including UNEP

INSTITUTIONS FOR PROTECTING THE ENVIRONMENT

and UNDP, should be adapted to changed circumstances in order to support sustained development;

among the goals of institutional reform at the global and regional levels should be an enhancement of the capacity of institutions at the national level, especially in developing countries, to ensure the full integration of environment and development;

states should promote and support, in more concrete ways, the effective participation of developing countries in the negotiation and implementation of international agreements or instruments—such support should include assured financial assistance to cover the necessary travel expenses to meetings and access to the necessary information and scientific-technical expertise on preferential terms;

ensure effective implementation and compliance, regular assessment and time review and adjustment or agreements or instruments by the parties concerned . . . and monitor and review domestic legal instruments giving effect to international agreements on environment and development, with a view to promoting compliance and harmonization;

improve plans and capability to respond to emergencies;

improve the effectiveness of institutions, mechanisms and procedures for administration of international environmental agreements and instruments;

assist developing countries and economies in transition in their national efforts at modernizing and strengthening the legal and regulatory framework of governance for sustainable development, having regard to local social values and infrastructures; and

disseminate information on effective legal and regulatory innovations in the field of environment and development, including noncoercive instruments and compliance incentives, with a view to encouraging their wider use and adaptation at the national and local level.[34]

The report called for "nothing less than an eco-industrial revolution" that will "create a whole new generation of economic opportunity and reduce the gross imbalance between the rich and the poor" and provide for effective global regulation:[35]

> The concept of national sovereignty has been an immutable and a most sacred principle of international law. There is no need for wholesale retreat from the principle. The present challenge is for us to recognise that, in many fields, particularly in the environmental field, it is simply not feasible for sovereignty to be exercised unilaterally by individual States, however powerful. Global environmental and economic security requires global cooperation which implies the willingness of States to give up, in appropriate legal instruments, a little of their sovereignty, in the interest of the international community at large.[36]

Effective enactment and enforcement of laws and regulations by governments and subgovernments are essential:

> The survey of existing agreements undertaken in the context of UNCED preparations has indicated serious problems of compliance in this respect, and the need for improved monitoring and related technical assistance.... There also needs to be an understanding on the part of the countries of the North that they will not divert the funds already earmarked for developments in the South and use such funds for dealing with newly identified environmental problems. Additional resources must thus be made available for dealing with environmental issues.[37]

Despite these recommendations, international institutions can rarely impose their will on sovereign states. The creation of a supranational environmental agency with powers to direct nations' behavior is not likely, at least in the short run. Much more realistic is the expectation that international organizations can encourage and facilitate political change in national and subnational governments rather than imposing it externally. Institutions can contribute to the formulation of agreements by supporting scientific research and disseminating the results, focusing attention on environmental issues and generating political pressure on governments to take action, placing environmental issues on the international policy-making agenda, providing a forum for negotiations, and linking environmental problems with other concerns.

Despite the shortcomings of existing environmental institutions, they will continue to perform important functions. Although the UN and affiliate organizations will continue to host conferences and negotiations, one of the most important tasks they and the more developed states can accomplish is to strengthen the institutional capacity of the other states. International organizations can do much without directly threatening the sovereignty of states.

Providing Financial and Technological Assistance

Without the support of international institutions, LDCs may not be able to participate effectively in negotiations, and their concerns may not be reflected in deliberations and policy formulation. They in turn will likely be more willing to comply with agreements that they believe take their concerns into account. Despite all this, during the UNCED meeting, the more developed nations agreed to make only very modest increases in their contributions to existing environmental and developmental efforts. The United States in particular has been reluctant to transfer new resources to LDCs. As a consequence, technical rather than financial assistance is likely to become an increasingly important source of help to LDCs.[39]

Some progress was made in the UNCED meeting in providing for the transfer of technologies from the MDCs to the LDCs. Under the Convention on Biological Diversity, for example, the parties agreed to transfer technology to developing countries under "fair and most favourable terms," while recognizing the "effective protection of intellectual property rights." Developing countries with genetic resources are to be "provided access to and transfer of technology which makes use of those resources, on mutually agreed terms, including technology protected by patents and other intellectual property rights" (Article 16). The United States was the only country that refused to sign the convention—an action that is hard to reconcile with basic notions of fairness and that appears quite shortsighted.[40] International organizations can make an important contribution by working out these differences so that LDCs can share in the benefits of harvesting their natural resources.

The Montréal Protocol on Substances that Deplete the Ozone Layer combines international regulations with financial incentives to help LDCs reduce their use of CFCs and replace them with safer substances. A $240 million fund was created in 1990 at the second annual meeting of the participating parties to help the LDCs that signed the protocol.[41] Debt-for-nature swaps between international nonprofit organizations and countries have reduced developing country debt with commercial banks by a very small amount, but do provide some protection for important natural resources (see Chapter 6). The size of the Third World debt (over $1.3 trillion) and the resultant interest payments means that in recent years there has been a net flow of funds from the South to the North. In 1990 the net flow was more than $30 billion, dwarfing the amount of debt reduced through swaps (about $100 million by 1990) as well as the assistance program provided for in the Montréal agreement.[42]

THE GLOBAL ENVIRONMENTAL FUND

The Global Environemntal Fund (GEF) was created to channel funds to the developing countries to help them implement programs that would reduce pressure on global ecosystems. The GEF is managed by the World Bank, the United Nations Environment Program, and the United Nations Development Program. Founded in 1990, the GEF has more than $1 billion to spend on projects aimed at global warming, biodiversity, international waters, and the ozone layer. The managing agencies are charged with ensuring that the facility enhances long-term, sustainable development. To qualify, a nation must have a per capita gross domestic product of less than $4,000. GEF funds are for additional programs; they do not replace existing development funds. Funds are limited to projects that foster implementation of treaty commitments, rather than projects aimed at general environmental concerns. Funds are also to be used to assist developing countries comply with the 1987 Montréal Protocol, the 1992 UNCED biological diversity and climate change conventions, and agreements to protect intrernational waters from pollution (the international Convention for the prevention of Pollution from Ships, the United Nations Convention on the Law of the Sea, the Protection and Preservation of the Marine Environment, and other agreements).[43]

As of 1995, fifty-three projects were earmarked for $470 million in funding, ranging from $4.8 million to monitor greenhouse gas emission to $10 million for biodiversity protection in East Africa to $7 million for a biomass gasification project in Brazil. Other projects include $10 million to create a trust fund in Bhutan, the proceeds of which will be used to hire foresters and others to protect biodiversity in the Himalayas; $10 million to fund the purchase of compact fluorescent lamps to reduce energy consumption and air pollution in Mexico; and the development of clean coal power technologies.[44] A small grants program gives up to $50,000 for community environmental projects and $150,000 for regional ones. The World Bank administers the trust fund; UNEP provides technical and scientific assistance in identifying and selecting projects, relying on an international Scientific and Technical Advisory Panel; and UNDP coordinates the financing and managing of technical assistance and preproject preparations. Both UNEP and UNDP help governments coordinate these projects with other efforts.

The Framework Convention on Climate Change and the Convention on Biological Diversity both designate the GEF as responsible for disbursing funds to help LDCs meet their obligations. Agenda 21 calls for a variety of funding mechanisms, including the GEF. In these accords, the developed nations agree to fund the "full incremental costs" of implementing the accords in the less-developed world; the developing countries are merely expected to comply with the mandates of the convention as long as financial and technical assistance is provided.[45] The treaties recognize that "economic and social developments are the first and overriding priorities of the developing country parties." The GEF plays a key role in the implementation of these conventions.

The climate change and biodiversity conventions stipulate that the GEF must meet several requirements:

Building Institutional Capacity for Environmental Protection

The GEF must operate within one or more existing international institutions. It is currently a part of the World Bank, but, as discussed above, the Bank has been widely criticized for failing to integrate environmental protection with its development efforts.

The Fund must involve the participating nations in decisions; although the conventions provide that each nation will receive one vote, thus giving power to the recipient states, donor states have insisted that they have a major role in determining how funds are spent.

Decisions must be open and subject to public scrutiny; host governments and NGOs will need to have access to information about projects.

During the initial stage, projects are primarily expected to generate ideas for innovative policy actions; eventually the Fund will need to focus on cost-effective projects. The atmospheric and biodiversity goals of Agenda 21 are estimated to cost more than $20 billion; although Agenda 21 goes well beyond the scope of the conventions, and the Fund is aimed at incremental costs, it will still require much more than the projected budget calls for in order to implement fully the agreements.

Funds must not be used to impose more conditions than are already placed on their use under the conventions.[46]

Projects may be funded by these three agencies (UNDP, UNEP, and the World Bank) as well as by regional development banks (such as the Inter-American, African, and Asian Development Banks, and specialized UN agencies such as the Food and Agricultural Organization, the International Maritime Organization, and the World Health Organization).[47]

To receive funds, a project must meet the following criteria:

1. The project must have sufficient legal and administrative structure and authority to support the project.

2. The project must have this additional funding for completion.
3. The project will result in global environmental benefits.
4. The project must be innovative—it must employ new technologies or ideas.
5. The project must be replicable, so what is learned can be applied elsewhere.
6. The project must contribute to one or more of the overall goals of reducing greenhouse gases, conserving biodiversity, protecting the ozone layer, or preventing or reducing ocean pollution.
7. The project must be consistent with sustainable development.
8. The project's effectiveness must be monitored and evaluated.
9. International organizations, NGOs, private companies, and national governments can all propose projects, but the governments must give final approval.

The 1991–94 pilot phase called for a total funding of $862 million, distributed as follows:[48]

Africa	20 percent of total spending
Asia	33 percent
Arab States and Europe	19 percent
Latin America/Caribbean	22 percent
Global	6 percent
Biodiversity	42 percent of total spending
Global warming	40 percent
International waters	17 percent
Ozone	1 percent

The GEF is an important experiment, and its success is central to the success of the conventions. Even if it is funded at the $2 to 4 billion level, however, it may not be enough to ensure effective implementation of the conventions. Future agreements will likely be integrated with the fund, but unless additional funds are provided, it will not be sufficient to ensure that developing countries will have the resources to participate in global agreements. Implementation is the key; will the projects actually work? Although the GEF is small, it may serve as a catalyst for other funding bodies, such as the World Bank, to consider environmental issues. Perhaps the greatest challenge is the different priorities of the North and South. For the North, protection of the global environment and natural resources is paramount; for the South, improving the quality of life for those who suffer from polluted air and water and from malnutrition and disease.

The GEF has been criticized for failing to address the development concerns of the South. The Fund was to come from new contributions, but critics argue that some of the money has simply been relabeled. The process has become complicated. For example, when projects are approved to aid countries in complying with the Montréal Protocol, UNEP transfers funds from the Montréal Protocol Multilateral Fund to the GEF's Ozone Projects Trust Fund, and the World Bank actually directs the expenditure of the money. Critics argue that the LDCs should have a greater say in how these funds are spent, but contributor nations resist relinquishing control over the funds they provide. Both UNCED conventions provided that funding mechanisms were to be accountable to all parties to the agreements and were to include an equitable and balanced system of representation and governance.[49]

Coordinating Efforts

One of the UN's most important roles is to coordinate environmental protection and international development efforts. Then Secretary General Boutros Boutros-Ghali, in a 1993 Conference on Global Development Cooperation, asked if the world had grown weary of development:

> Many donors are tired. Many of the poor are dispirited. The field of development cooperation is in crisis. . . . We look back on a record of frustration. The rich countries had hoped to

resolve the problems of the South—poverty, hunger, high population growth rates, low productivity, stagnation, debt, drought, desertification—by channeling aid through international and national institutions which were created almost 50 years ago. This has however not been a success. The time has come to take a critical look at the aid patterns of the past.[50]

The key problem was coordination: "Global development cooperation is obstructed when national and international agencies number in the hundreds and each goes its own way." One solution is for bilateral and multilateral donors "to complement efforts of developing countries at national macro-management. They can do so by providing resources that will allow these countries to build up their social and physical infrastructure." The UN should "facilitate inter-agency coordination as well as coordination with non-governmental organizations." Boutros-Ghali endorsed regional assistance programs in particular:

> Regional aid may be preferable to bilateral aid in many areas. Regional aid follows practical needs wherever they may go, and does not stop at political boundaries. . . . Regional organizations would be reinforced—and recipients would benefit—if more aid came from, or through, organizations at the regional level. This would allow recipient countries to deal with one donor rather than many. Such regional efforts often are more able to design an effective division of labour in the field of aid. From there the regional trade-offs can be more clearly seen. And the bureaucratic competition of the donors may be transcended.

The former secretary general concluded with a somber warning: "The United Nations has a unique role to play in global development cooperation, provided that it takes up the double challenge of changing its thinking and its structure."[51]

The challenge of reorienting UN activities is daunting. One of the key roles of the United Nations, for example, has been to collect and disseminate information concerning global trends in health, agricultural output, energy, education, population, trade, and so

Building Institutional Capacity for Environmental Protection

forth. Data collection has decreased over time as budgets have failed to keep up with growing programmatic responsibilities. The 1972 Stockholm Conference called for the creation of a new agency to monitor environmental trends. UNEP, which assumed that responsibility, has funded efforts to monitor environmental conditions, conduct research, and exchange information. In 1974 it established Earthwatch, an informal alliance of the World Meteorological Organization's Global Atmosphere Watch. UNEP also sponsored tropical forest assessments by the Food and Agriculture Organization and water quality monitoring by the World Health Organization and the WMO. The UN Statistical Office provides standards for collection and publication of data by member states; the UN Advisory Committee for the Coordination of Information Systems issues guidelines for environmental data; and the Office for Research and Collection of Information collects information about member states for the secretary general. Overlap, duplication, and lack of coordination hamper even information collection.[52]

Institution Building

International organizations, one study concluded, can empower "domestic actors to solve domestic problems of international importance." They can "foster capacity building by providing policy-relevant information that can be used by government allies to develop better programs and to justify their actions to domestic opponents."[53] The UN Sustainable Development Commission, charged with monitoring compliance with the agreements signed at the Rio Summit, has the potential to contribute to these efforts.

One of the most important functions of these institutions is to disseminate information that others can use to generate political support for preventive and other actions. Another study concluded that:

> the most significant roles of international institutions—such as magnifying concern, facilitating agreement, and building capacity—do not require large administrative bureaucracies.

Indeed, running such a bureaucracy may divert leaders of international organizations from their most important tasks, which are quintessentially political: to create and manipulate dynamic processes by which governments change conceptions of their interests and to mobilize and coordinate complex policy networks involving governments, NGOs, subunits of governments, and industrial groups, as well as a variety of international organizations that have different priorities and political styles.[54]

Information concerning alternative policy tools, such as decentralized, marketlike regulatory approaches, forms another category of information that could be fruitfully disseminated.[55]

The Importance of Local Efforts

Nongovernment organizations are enormously important to the implementation of global environmental accords. NGOs are tremendously varied. Some are grassroots organizations that work in villages and urban communities in the less-developed nations. Others are service organizations that provide support for grassroots groups. Some NGOs have grown into regional or national federations, and some form international networks and coalitions. Many NGOs focus on particular issues such as development, the environment, human rights, relief, or women's issues; others provide services such as policy research, financial and technical assistance, and training to grassroots groups.[56]

NGOs mobilize collective action and render important economic and other services such as providing sources of credit, access to marketing services, technical support, educational programs, and health care. NGOs give voice to grassroots concerns and become effective political and economic forces. Studies of NGOs involved in development have found them to be "relatively effective at reaching the poor, mobilizing local resources, delivering services, and solving problems." NGOs have generally been better able to help poor individuals who have some resources rather than those who have few or no assets. Their "administrative flexibility, small

Building Institutional Capacity for Environmental Protection

size, and relative freedom from political constraints make it easier for successful service organizations to try innovative solutions to problems."[57]

Grassroots organizations are invaluable in mobilizing local energy to address pressing problems. They can pressure government officials to support international agreements and pursue additional domestic initiatives. NGOs expand the range of interests that are included in policy making and play a particularly important role in ensuring that local concerns and interests are reflected in policy deliberations, increasing the likelihood that programs will be sustained over the long run. NGOs can integrate into public policies and community efforts concerns that are relevant to development, environmental quality, public health, literacy, justice, and culture. National, regional, and international NGOs can spread knowledge and help compensate for differences in the distribution of wealth, resources, and technologies.[58]

Although NGOs have played and will continue to play a major role in environmental cleanup and preservation efforts, the experience has been mixed and many groups have failed. Some service NGOs have been criticized for failing to reach the poorest in less-developed nations, for becoming as inflexible as traditional governmental bureaucracies, for lacking the necessary technical expertise and organizational skills, for being unable to replicate successful projects, and for not sustaining their efforts after the withdrawal of initial assistance from outside groups. NGO efforts often lack long-term planning, largely because of the instability of funding. Rather than coordinating their efforts, NGOs that operate in the same geographic areas often compete for scarce resources, government support, and local allegiances. There is usually little coordination with, and sometimes conflict between, NGO and government programs. Some NGOs fail to invest in the development of administrative infrastructure and the institutional capacity building necessary for the long-term viability of their efforts. Local grassroots organizations may find it difficult to expand their activities or influence broader governmental efforts. Technical expertise is often scarce. Efforts that remedy one problem, such as reducing emissions from one local source, may not be able to prevent other problems, such as pollution from other areas.[59]

One of the most promising developments is the creation of network organizations, which bring together NGOs to exchange information and coordinate activities. One such organization is the Indonesian Environmental Forum, made up of more than four hundred smaller environmental organizations, which organizes conservation education programs, sponsors training programs on assessing environmental impacts or environmentally appropriate technology, provides technical assistance for issues such as fund raising, lobbying government officials, and bringing lawsuits, including Indonesia's first suit against a corporation for violating environmental laws.[60]

Institutions as Catalysts

If we are to protect the global environment, we must create new global bodies or at least reinvigorate existing ones. But global government that can override national sovereignty cannot bring about the changes required to respond to these threats. Although environmental preservation is becoming increasingly important, national security, economic competitiveness, and other policy demands may be given higher priority. Implementation must occur in so many different places, where environmental conditions and institutional, political, economic, cultural, and social practices differ so widely that a centralized approach is futile. Despite the considerable barriers to making the UN a more effective force for implementing global environmental agreements, even greater constraints lie in the nature of the problems to be addressed and the kinds of changes that are required. The immediate challenge is to strengthen all nations' institutional capacity to fashion effective responses to the challenges of environmental degradation.

Solutions to environmental problems require fundamental changes in our ways of thinking and doing, changes that are for the most part beyond international bureaucracies. Perhaps the greatest role that international organizations may play in the implementation of global

References

agreements will be an indirect one. These organizations can help participating nations with their implementation plans, by monitoring their progress and facilitating the sharing of scientific and regulatory information. However, the gap between global formulation and local implementation is enormous. The real public policy challenge is to devise ways for all levels of governments to be effective catalysts for change in individual behavior.

References

1. United Nations, *Basic Facts About the United Nations* (United Nations, 1992), 3, 7, 260.
2. Ibid., 7–11.
3. Ibid., 273.
4. See Harold Stassen, *United Nations: A Working Paper for Restructuring* (Minneapolis: Lerner Publications, 1994).
5. Lynton Keith Caldwell, *International Environmental Policy*, 2d ed. (Durham, N.C.: Duke University Press, 1990), 282–86.
6. The source of the listings of UN organizations is Lee Kimball, *Forging Environmental Agreement* (Washington, D.C.: World Resources Institute, 1992), 70.
7. Thomas F. Malone and Robert W. Corell, "Mission to Planet Earth Revisited," *Environment* (April 1989).
8. Jyoti Shankar Singh, "How to Reduce the Debt Burden," *Earthtimes* (1–15 October 1996): 12.
9. See Bruce Rich, *Mortgaging the Earth: The World Bank, Environmental Impoverishment, and the Crisis of Development* (Boston: Beacon Press, 1994); Kevin Danaher, ed., *50 Years is Enough: The Case Against the World Bank and the International Monetary Fund* (Boston: South End Press, 1994).
10. Pratap Chatterjee and Matthias Finger, *The Earth Brokers: Power, Politics, and World Development* (London: Routledge, 1994); Johan Holmberg, ed., *Making Development Sustainable: Redefining Institutions, Policy, and Economics* (Washington, D.C.: Island Press, 1992).
11. Richard W. Stevenson, "World Bank Chief Asks Slimmer Staffs and Better Lending," *New York Times*, 21 February 1997, C1.
12. Elizabeth Bryant, "The Bank and The Fund," *Earthtimes* (1–15 October 1996): 7–8.

13. World Resources Institute, *World Resources 1994–95* (Washington, D.C.: WRI, 1994), 224.
14. Ibid., 224–25.
15. Ibid., 225.
16. World Resources Institute, *World Resources 1992–93* (New York: Oxford University Press, 1992), 68–69.
17. Ibid., 68–69.
18. Kimball, *Forging International Agreement*, 69.
19. World Resources Institute, *World Resources 1992–93*, 68–69.
20. World Resources Institute, *World Resources 1994–95* (Washington, D.C.: World Resources Institute, 1994), 225.
21. *Common Responsibility in the 1990s: The Stockholm Initiative on Global Security and Governance* (Stockholm, 1991).
22. See, generally, Jan Knippers Black, *Development in Theory and Practice* (Boulder, Colo.: Westview Press, 1991); H. Jeffrey Leonard, *Environment and the Poor: Development Strategies for a Common Agenda* (New Brunswick, N.J.: Transaction Books, 1989); and Robert J. Berg and David F. Gordon, eds., *Cooperation for International Development: The United States and the Third World in the 1990s* (Boulder, Colo.: Lynne Reiner, 1989).
23. Judith T. Kildow, "The Earth Summit: We Need More Than a Message," *Environmental Science and Technology* 26 (June 1992): 1077.
24. Paul Lewis, "Peacekeeper Is Now Peacemaker: UN Wrestles with Its New Roles," *New York Times*, 25 January 1993.
25. See Michael R. Gordon and John H. Cushman, Jr., "After Supporting Hunt for Aidid U.S. Is Blaming UN for Losses," *New York Times*, 18 October 1993.
26. Paul Lewis, "United Nations Is Finding Its Plate Increasingly Full But Its Cupboard is Bare," *New York Times*, 27 September 1993.
27. Daniel J. Shepard, "Renewed Efforts at Reforming UN," *Earthtimes* (1–15 October 1996): 5.
28. "UN Must Reduce Dependence on U.S. Funding, leader says," *Deseret News*, 14 August 1995.
29. Leonard Silk, "Don't Strangle the UN," *New York Times*, 9 October 1993.
30. Marc A. Levy, Peter M. Haas, and Robert O. Keohane, "Institutions for the Earth: Promoting International Environmental Protection," *Environment* 34 (May 1992): 33.
31. Ibid.
32. Ibid., at 32.
33. World Resources Institute, *World Resources 1994–95*, 226.

References

34. A. O. Adede, "International Environmental Law from Stockholm to Rio—An Overview of Past Lessons and Future Challenges," *Environmental Policy and Law* (22/2 1982): 100–101.
35. Ibid., 101.
36. Ibid., 102.
37. Ibid., 102.
38. *Common Responsibility in the 1990s: The Stockholm Initiative on Global Security and Governance* (22 April 1991): 43–45.
39. Adede, "International Environmental Law from Stockholm to Rio," 88–105. See also A. O. Adede, "A Profile of Legal Instruments for International Responses to Problems of Environment and Development," *Environmental Policy and Law* (21/5–6 1991): 224–32.
40. Andy Coghlan, "Biodiversity Convention a 'Lousy Deal,' Says US," *New Scientist* (4 July 1992): 9; "Bush 'Shortsighted,'" *New Scientist* (11 July 1992): 9.
41. See Richard Elliot Benedick, *Ozone Diplomacy* (Cambridge, Mass.: Harvard University Press, 1991).
42. Morris Miller, *Debt and the Environment* (New York: United Nations Publications, 1991).
43. John C. Dernbach, "The Global Environment Facility: Financing the Treaty Obligations of Developing Nations," *Environmental Law Review*, 23 (March 1993): 10124.
44. Dernbach, "The Global Environment Facility," 10126.
45. *Framework Convention on Climate Change*, art 4, paras. 3,7; *Convention on Biological Diversity*, art. 20, paras 2,4.
46. *FCCC*, arts. 11, 21; *CBD*, arts. 21, 39; Dernbach, "The Global Environment Facility," 10128–30.
47. Dernbach, "The Global Environment Facility," 10125–26.
48. World Resources Institute, *World Resources 1994–95*, 230.
49. Ibid.
50. Boutros Boutros-Ghali, address to "Conference for Global Development Cooperation," Carter Center of Emory University (4–5 December 1992).
51. Ibid.
52. Kimball, *Forging International Agreement*, 15.
53. Levy, Haas, and Keohane, "Institutions for the Earth," 34.
54. Notes from Editors of the Harvard Law Review, *Trends in International Environmental Law*, (Chicago: American Bar Association, 1992), 32.
55. For an introduction to this literature, see Walter Rosenbaum, *Environmental Politics and Policy* (Washington, D.C.: Congressional Quarterly Press, 1993).

56. For an overview of NGOs, see World Resources Institute, *World Resources 1992–93*, 216–21.
57. Ibid., 221–22.
58. Ibid., 232.
59. Ibid., 230.
60. Ibid., 230.

5

The Political Economy of Global Environmental Regulation

THE NATIVE PEOPLES of the Americas have not always lived in harmony with nature. Ecological damage threatened some societies and brought about the extinction of others. One of the lessons we learn from the thousands of years of western history is the importance of living in harmony with the environment. The Maya, for example, may have experienced the rapid collapse of their society because of ecological destruction, deforestation, and unsustainable agricultural practices.[1] Nevertheless, native peoples have traditionally shared a strong commitment to their land and have sought to live in harmony with it. Their relationship goes beyond subsistence living to intimate, spiritual interconnectedness: their lands are sacred, manifestations of the Great Spirit, and places of profound spiritual importance.[2] Many believe that everything is connected to everything else, as is reflected in a widely quoted statement that has been attributed (although apparently inaccurately) to Chief Seattle:

> How can you buy or sell the sky? The land? The idea is strange to us. If we do not own the freshness of the air and the sparkle of the water, how can you buy them? Every part of this earth is

sacred to my people. . . . This we know: the earth does not belong to man, man belongs to the earth. All things are connected like the blood that unites us all. Man did not weave the web of life, he is merely a strand in it. Whatever he does to the web, he does to himself.[3]

The Hau de no Saunee taught that "in every deliberation, we must consider the impact of our decisions on the next seven generations." "When we burn grass for grasshoppers," remarked a Wintu woman, "we don't ruin things. We shake down acorns and pinenuts. But the White people plow up the ground, pull down the trees, kill everything. . . . They blast rocks and scatter them on the ground. . . . How can the spirit of the Earth like the White man? . . . Everywhere the White man has touched it, it is sore."[4] A chief of the Blackfeet, in response to a treaty he was asked to sign, said:

> Our land is more valuable than your money. It will last forever. It will not even perish by the flames of fire. As long as the sun shines and the waters flow, this land will be here to give life to men and animals. We cannot sell the lives of men and animals; therefore we cannot sell this land. It was put here for us by the Great Spirit and we cannot sell it because it does not belong to us. You can count your money and burn it within the nod of a buffalo's head, but only the Great Spirit can count the grains of sand and the blades of grass of these plains. As a present to you, we will give you anything we have that you can take with you; but the land, never.[5]

For Begadi, or Big Thunder, of the Wabanakis Nation, the "great spirit is our father, but the earth is our mother. She nourishes us; that which we put into the ground she returns to us and healing plants she gives us likewise. If we are wounded, we go to our mother and seek to lay the wounded part against her, to be healed."[6] Tatanga Mani, a Stoney Indian, wrote in his autobiography:

> Oh, yes, I went to the white man's schools. I learned to read from school books, newspapers, and the Bible. But in time I found that these were not enough. Civilized people depend too

Interaction of Environmental Protection and Economic Growth

much on man-made printed pages. I turn to the Great Spirit's book which is the whole of his creation. You can read a big part of that book if you study nature. You know, if you take all your books and lay them out under the sun, and let snow and rain and insects work on them for a while, there will be nothing left. But the Great Spirit has provided you and me with an opportunity for study in nature's university, the forests, the rivers, the mountains, and the animals which include us.[7]

For five hundred years the Western world has been guided by ideas of growth and progress that have come to profoundly threaten our survival. A sustainable ecosystem—one that provides a hospitable climate for current and future life—is a fundamental challenge for public policy making..

The Interaction of Environmental Protection and Economic Growth

Some writers have argued that a transformation to an ecologically based commerce is about to occur, that businesses are on the verge of reconstructing commerce, or that economic incentives have become sufficiently powerful to ensure that pollution will eventually be eliminated and environmental goals achieved. As discussed in Chapter 3, market-based approaches may help reduce the conflict surrounding environmental regulation. However, optimism about the role of economic theory in accomplishing global environmental goals must be tempered by an examination of the political challenges involved in integrating economic theory into the formation and implementation of global environmental agreements.

From one perspective, economic and environmental goals are incompatible. The global environment will not likely be able to sustain the level of consumption that residents of the more developed world now enjoy. The increasing demands for resources, combined with population growth and the production of pollution, are a recipe for disaster. Scholars and policy advocates call for a dramatic shift in our ways of thinking and action and for a transformation to a new way of interacting with the environment.[8]

In the short run, effective environmental policies will often be disruptive. They will challenge the existing distribution of jobs and profits and will create new economic winners and losers. They will require expensive investments that will take years to amortize. Many costs will be immediate and most benefits will not be realized until far into the future. We may need to sacrifice to short-run economic considerations ensure a sustainable environment for current and future generations. In the long run, however, others argue, most environmental and economic goals are compatible. Modernization of industrial processes can result in improved product quality and reduced emissions. Improved energy efficiency reduces costs and environmental damage. Environmental regulation results in money spent on pollution control equipment and jobs and growth in the fledgling pollution control industry. A growing economy will generate the resources necessary to protect the environment.[9]

The interaction of economic and environmental goals and concerns is a fundamental consideration in the implementation of global environmental agreements. Sometimes the debate is relatively narrow: Environmental advocates argue that the benefits promised from regulation outweigh projected costs, whereas industry representatives dismiss these benefits as falling considerably short of the burdens to be borne. Some argue that costs are irrelevant, that human health must be protected regardless of costs, or that benefits are so difficult to estimate that any comparison of costs and benefits is unfair. Others claim that environmental regulations pose serious barriers to the global competitive position of U.S. industry and that stringent protection will simply drive jobs overseas. The question has often been posed in stark terms: Should we give priority to the protection of human health and ecological systems, or to economic growth and competitiveness?[10]

Economic considerations are obviously critical in determining whether regulatory initiatives will be pursued. But even where the benefits and costs within national economies will produce some winners and some losers, the redistribution of benefits and burdens will produce corresponding political changes. In the international political economy, developing countries face far more immediate

Interaction of Environmental Protection and Economic Growth

and critical pressures to stimulate economic growth and remedy poverty than the developed nations, and have many fewer resources available to do so. These countries may oppose proposals by the developed countries to reduce emissions or adopt new technologies that will restrict opportunities for development. A related concern is that, given the tremendous capital requirements that might be required to address threats like climate change, it is essential to pursue policies that ensure a strong and growing economy. The greater the economic growth, the more resources will be available to invest in needed technologies and fund necessary adaptations.[11]

From a traditional, neoclassical economic perspective, environmental regulation may still encounter strong resistance, despite arguments that it is consistent with a market economy. Environmental regulation can be viewed as being quite compatible with and even necessary for a free market economy if it seeks to ensure that the true costs of production are included in the prices charged (rather than being externalized or imposed on third parties). Some critics argue that advocates of aggressive regulation are out to destroy capitalism.[12] In one sense, that may be true; some may argue that capitalism, with its emphasis on private property rights, consumption, pursuit of profits, and immediate fulfillment at the expense of long-term considerations, is the problem. But other champions of regulation argue that markets are not the problem; in fact, making markets work is the solution. Regulation is aimed at eliminating externalities; the goal is to foster markets that include true costs.[13]

Skeptics respond, however, that regulations are not usually so neatly devised. Decisions about the allocation of resources, production, consumption, and so on, they maintain, should be made through markets, not through political means. Regulation involves high bureaucratic or transaction costs, the loss of individual freedom, the stifling of innovation, and the creation of opportunities by businesses to use government regulation as a way to gain some advantage over competitors (rent seeking, as economists call it). The benefits of environmental protection, they argue, are tenuous and uncertain, whereas restrictions on business and personal freedom are certain.[14]

Skeptics of regulation also argue that being richer is healthier and safer than being poor. Regulation raises prices and hinders economic activity, eventually making us less safe because we have fewer resources to respond to problems. Profitable companies will have more resources to modernize and thus produce less pollution. If we foster wealth-generating behavior now, then when we are more certain that environmental problems exist, we will sufficient resources to respond. Similarly, opportunity costs are significant: We can spend our limited resources more profitably by addressing definite needs rather than hypothetical environmental catastrophes.[15] In the case of global warming, for example, such critics are likely to argue that since much uncertainty exists, we should wait to commit resources.

Two areas—international trade and technology—raise many of the issues that are central in implementing global environmental agreements and in exploring the possibilities of integrating economic and environmental goals.

Rethinking International Trade

International trade agreements seek to standardize trade rules to allow products to be traded more freely. Multinational corporations are the primary interests that support this version of "free trade," since they stand to benefit the most. These agreements are better characterized as managed trade rather than free trade, since numerous provisions are aimed at helping specific industries through subsidies and tariffs for industries that are represented in the negotiations. Most problematic are provisions that countries cannot discriminate on the basis of the method of production unless scientific justification exists. However, an international body determines whether sufficient scientific basis exists for more stringent domestic regulations. As a result, environmental laws are subordinated to international trade goals.

Free trade can be compatible with environmental quality, but if markets are not working well—if they are not taking into account all the costs of production—the advantages of free trade and markets will not be realized.[16] We need to encourage companies that are

Interaction of Environmental Protection and Economic Growth

trying to restore the environment and contribute to a sustainable economy; they will not make the attempt if they are competing against companies that are unwilling to include true costs in their prices. We need to encourage companies to engage in production in ways that reduce consumption, energy use, distribution costs, and environmental damage. Trade agreements, if they continue to permit externalizing of costs, are little more than ways for multinationals to expand their profits.[17]

For two centuries the modern, industrialized world was dominated by trade between a small number of industrialized nations and the rest of the world, which provided raw materials. That has all changed with the rise of newly industrialized nations and multinational corporations in the late twentieth century.[18] Underlying this shift in industrial structure is the fear that aggressive environmental regulation in the more-developed nations would encourage industry to find havens in the less-developed world where there is less commitment to environmental regulation. Economic disruptions and increased pollution would be the result. Rather than fleeing to pollution havens, however, most industries have developed new technologies and changed production processes in order to comply with environmental standards. Other factors such as market competition, transportation infrastructure, the availability of raw materials, labor supply, and political support have been more important than regulatory concerns. Although some countries have tried to gain competitive advantage by neglecting environmental protection, and some firms have relocated their facilities to nations with weaker environmental standards, the shift in manufacturing has occurred largely in response to economic, not environmental demands.

Support for free trade appears to be one of the most widely held views in the world today. It promises economic efficiency, growth, technology transfer, and improved quality of life for all global citizens. But as competitive forces grow, so does the pressure to externalize environmental costs, rather than reflect them in true or real prices. As international competitiveness becomes increasingly important, environmental quality may be increasingly sacrificed in order to reduce costs.

Free trade rests on a number of important assumptions and values. Countries should specialize in what they can produce more efficiently than others—where they have a comparative advantage. They can then trade for the goods and services they seek to use and consume. Specialization and comparative advantage require several things to happen. First, there should be no externalities: All the costs of production are included in prices. Second, prices should be stable over time: If countries see a comparative advantage in some goods but produce them at levels where prices fall below production costs, specialization will be disastrous. Third, comparative advantages are assumed to be constant, but in reality they may vary over time, and decisions about investments in human and natural resources can alter, over time, the advantages some countries enjoy Leaving these developments to markets may ultimately be a catastrophe. Fourth, decisions about consumption and production must not be coerced: Choices must be responses to economic incentives, but there are tremendous disparities in wealth and power within and among countries that shape these choices along other lines. Fifth, capital is expected to stay within national boundaries: If capital can flow to countries that enjoy comparative advantage, then other countries will lose the economic resources necessary to compete.[19] Since none of these preconditions prevail in current international trade, we ought to be skeptical about the claims made for free trade. There is no question that free trade benefits the already wealthy and powerful, but its environmental costs and other consequences for the developing world raise red flags.

More broadly, specialization and comparative advantage are part of a broader assumption that economic growth is limitless, that resources are freely substitutable, and that scarcity is naturally accommodated through market prices. But if environmental limits are central to the prospects for human survival, if some resources, such as clean air and water, cannot be freely substituted, then traditional economics and its reliance on free markets cannot serve as the basis for global decisions. We must make free trade be made compatible with environmental protection—and human rights, worker protection, community decision making, and the quality of life in the developing world.

Interaction of Environmental Protection and Economic Growth

THE GENERAL AGREEMENT ON TARIFFS AND TRADE

The General Agreement on Tariffs and Trade (GATT) is rooted in this vision of free trade and economic growth, which has become the overriding global concern of nations. GATT's primary goals are to reduce barriers to trade, foster foreign investment, and protect the property rights of foreign investors. GATT does not provide a framework that requires markets to take environmental and social costs into account. Recognition within countries is widespread that government is required to provide the framework in which market exchanges can occur; to ensure competition, fair trade, integrity of contracts; and to provide for transportation, communication, and other public goods. However, there is little recognition of the need for such frameworks for international trade. Trade does not allow countries to exclude products produced in environmentally harmful ways. GATT provides no opportunity for environmental groups to participate directly in proceedings; it created the Committee on Trade and the Environment (CTE) in 1994 to make recommendations in areas where trade and environmental protection may conflict. Governments may invite groups to work with them in CTE proceedings. The Clinton administration has expanded such opportunities for representatives from environmental groups, but other countries have not because of fears such efforts would inhibit freer trade.[20]

GATT permits parties to challenge the laws or regulatory programs of other countries that violate GATT rules concerning acceptable trade measures. These rules govern the kinds of import controls parties can place on foreign-made products and foreign production processes, the trade sanctions that can be imposed, and product and process standards applicable to imported and domestic goods. GATT rules require that products from other countries be treated no less favorably than domestic-produced products,[21] prohibit quantitative restrictions on imports, and provide for a number of justifications for trade restrictions for measures

> necessary to protect human, animal or plant life or health; relating to the conservation of exhaustible natural resources if such measures are made effective in conjunction with restrictions on domestic production or consumption; and necessary to secure

compliance with laws or regulations which are not inconsistent [with GATT].[22]

GATT panel decisions are not themselves binding, but are only recommendations to the GATT Council that it request defendant countries to bring their laws and policies into harmony with GATT standards. If the Council approves the recommendation but the defendants do not comply, plaintiffs can impose retaliatory trade sanctions. Under changes made in the Uruguay Round of GATT, panel reports are automatically approved unless the Council acts to reject them, and a three-fourths vote of members can waive the application of GATT provisions. The United States has been a defendant in GATT panel hearings more than any other nation, as well as being the most active plaintiff.[23]

The experience of the United States in protecting dolphins in the Pacific Ocean illustrates one kind of tension between free-trade agreements and environmental regulation. The 1972 U.S. Marine Mammal Protection Act authorized the Federal government to embargo tuna from other nations caught by any means that results in the incidental killing of or injury to ocean mammals beyond what is permitted under U.S. standards.[24] Tuna fishing boats often follow dolphins in order to locate the tuna that dolphins eat, and then cast nets on the dolphins to harvest the fish. Amendments enacted in 1984 required nations from which the United States imports tuna to demonstrate they have a dolphin conservation program comparable to the U.S. program.[25] Amendments in 1988 imposed a limit on foreign nations operating in the eastern tropical Pacific Ocean: Their dolphin take must be no more than 125 percent of what U.S. vessels took. Another amendment required the Federal government to impose an intermediary embargo on tuna from countries that import it from nations subject to a primary embargo by the United States.[26] The 1992 amendments (also called the International Dolphin Conservation Act) authorized the lifting of primary embargoes on countries that commit to a five-year moratorium on using nets on dolphins in catching tuna, and banned the sale of or importation into the United States of tuna that is not "dolphin safe"—requiring, among other things, that dolphin nets not be used in harvesting tuna.[27]

Interaction of Environmental Protection and Economic Growth

In September 1990 the U.S. Customs Service banned tuna imported from Mexico, Panama, and Ecuador. When Mexico challenged the ban in 1991, a GATT panel was convened to hear Mexico's claims. The United States acknowledged that the MMPA violated the GATT prohibition against quantitative limits on imports, but argued that it nevertheless satisfied GATT's standard on treating imported products no less favorably than domestically produced ones. The panel found the U.S. action violated GATT because it: (1) did not qualify as an internal regulation applying equally to imports and did not fall within any of the general exemptions; (2) constituted a quantitative restriction; (3) and was a unilateral action on the part of the United States.[28]

In 1992, the European Union and the Netherlands (on behalf of the Netherland Antilles) challenged the United States primary and intermediate embargo policies. In May 1994, the GATT dispute panel again ruled that the U.S. policies violated GATT. Some reasons were similar to the earlier ruling: Since U.S. law regulated harvesting methods or production processes that had no impact on tuna as a product, the law was inconsistent with GATT. The panel acknowledged that dolphins were an exhaustible natural resource and that countries could seek to protect such resources that lie outside their territorial jurisdiction, but it concluded that the prohibition of any tuna imports from intermediary countries could not further U.S. conservation goals, since some tuna might be harvested independent of any dolphins. It rejected the U.S. law as an effort to force other countries to change their domestic policies, an action inconsistent with the overall framework of GATT. Although the United States could seek to protect resources outside its jurisdiction, the panel ruled that such efforts must rely on measures that entail "the least degree of inconsistency with other GATT provisions," and the U.S. attempt to force other countries to change their policies failed that test.[29]

GATT panel decisions have posed numerous problems. The decisions argue that GATT gives members the "right" to use trade regulation to protect life or health, whereas that power is actually inherent in national sovereignty and was not surrendered when GATT was created in 1947. The decisions provide no evidence that one

225

country's efforts to force others to change their policies jeopardizes the entire GATT structure. They endanger the application of standards based on processes or production methods that many nations use to protect natural resources and the environment.[30] Although the panels claimed they were not assessing the validity of U.S. laws, their decisions undermined efforts to protect Pacific dolphins: Because GATT ruled that the United States can regulate only its own citizens, but is powerless to try to affect the behavior of others, as a result the United States will not be able to achieve its goal. Other criticisms focus on the nature of the panel process: The panels are closed to the public, experts are not invited to testify, and only governments have standing.[31]

More broadly, the underlying dispute here is an environmental and not a trade one: How much protection should be given dolphins, and who should pay? But GATT panels are not allowed to base their decisions on broader principles of international law or address issues beyond trade. Given the confusion over the interaction of GATT with other international agreements, forums for settling environmental conflicts are needed. Under the UN Law of the Sea treaty (UNCLOS), for example, a party could challenge a trade embargo and ask for a compulsory dispute settlement proceeding; regardless of what a GATT panel had said, the UNCLOS panel could rule that the disputed measure violated UNCLOS or that it did not and the losing party must reform its conservation policies. UNCLOS provisions indicate that its panels will defer to binding decisions resulting from dispute settlement procedures in other agreements, but some argue that GATT panel decisions are advisory and not binding, although that may no longer be true as a result of the Uruguay Round.[32]

Because of the dolphin case, opponents of GATT charged that U.S. environmental standards could be challenged as unfair trade barriers in the World Trade Organization. The final bill ratifying the latest GATT agreement included only a few environment-related provisos: Representatives from environmental groups must be included in the U.S. Advisory Committee for Trade Policy and Negotiations, and the WTO dispute settlement panels are to be opened up to more public participation and scrutiny.[33] Despite these efforts, there have been problems between the United States and the WTO during its first years. In January 1996, a dispute-settlement panel ruled that a U.S. regulation that gasoline be reformulated to meet standards under the Clean Air Act was a restraint of trade since the United States gave more time to domestic producers to meet the new standards than it gave to foreign sources. The panel ordered the United States to reformulate its air pollution laws and regulations.[34] The United States appealed the decision, and a WTO internal review board affirmed the ruling in April 1996. The United States can block adverse GATT rulings, but it is obligated to under the WTO agreement to change the regulation, face trade sanctions, or offer offsetting trade benefits. The decision gave ammunition to opponents of free trade who argued that the WTO would require the United States to weaken domestic environmental and health protections.[35] In February 1997, European countries challenged the United States in a World Trade Organization proceeding because of its embargo on Cuba. The United States imposes penalties on countries that do business in Cuba, such as prohibiting their executives from entering the United States. The United States refused to participate in the proceeding, claiming that the issue was one of national security and had nothing to do with trade. The United States brings more cases to the WTO than are brought against it, and the rejection of the organization's jurisdiction in this dispute may jeopardize future cases.[36]

THE NORTH AMERICAN FREE TRADE AGREEMENT

Like GATT, the North American Free Trade Agreement (NAFTA) is primarily an effort to reduce trade barriers, and only offers some procedural protection for the environment. In the 1980s, as efforts to negotiate more free international trade stalled, attention shifted to developing regional trade agreements such as NAFTA and Mercosur, a trading pact among South American countries. These regional trade associations were seen as the best available option for liberalizing trade and were expected to encourage opponents of free trade, such as Japan and France, to support open markets. But some studies have shown that these arguments adversely affect those outside the pacts; some economists argue that regional agreements are inefficient and complicate trade as well as limiting access to goods and services. In reality, regional pacts have not been created to promote free trade, but seek other purposes. The most developed regional agreement, the European Union, began as a way to strengthen opposition to communism. NAFTA was championed as a way to encourage Mexican modernization and political stability.[37] Regional trade agreements may restrict global trade, but may also provide an opportunity for collective efforts to pursue environmental goals as has happened in the European Union and may happen in NAFTA. But there are numerous critics of the environmental consequences of these agreements.

The North American Agreement on Environmental Cooperation (NAAEC) created two public advisory bodies, one to advise the Committee on Environmental Cooperation (CEC), and a second to assist the Border Environmental Cooperation Committee (BECC). These two commissions are to monitor the environmental conditions affected by increased trade. Although the NAAEC encourages nations to improve food safety and technical standards, and to not reduce environmental standards in order to attract investment, it gives little recognition to the importance of internalizing environmental costs and the challenges to less-developed countries in competing in global markets without further compromising environmental quality and natural resources.

The debate over the North American Free Trade Agreement during the fall of 1993 demonstrates how environmental concerns might be integrated with economic ones. Many environmentalists opposed the treaty, warning that it was an attempt by polluting industries to reduce environmental quality standards under the guise of lowering trade barriers. However, considerable efforts went into respond to these concerns. NAFTA provides that each party may set its "appropriate level of protection" for human health and ecosystems, with two limits: (1) parties "shall, with the objective of achieving consistency in such levels, avoid arbitrary or unjustifiable distinctions in such levels in different circumstances . . . where such

Interaction of Environmental Protection and Economic Growth

distinctions result in arbitrary or unjustifiable discrimination against a good of another Party or constitute a disguised restriction on trade;"[38] and (2) when regulating animal or plant pests or disease, parties are to consider "the cost-effectiveness of alternative approaches to limiting risk."[39] Trade measures aimed at protecting health and environment must be limited to actions that are "necessary" to protect life or health.[40] NAFTA also provides that trade measures be "based on scientific principles" and eliminated if that scientific base no longer exists.[41] However, since many have questioned the scientific basis of U.S. domestic regulations, these measures could be challenged.

NAFTA may turn out to be more amenable to environmental protection than GATT. The Uruguay Round of GATT requires that "Sanitary and Phytosanitary Measures" be the "least restrictive to trade, taking into account technical and economic feasibility," a much more difficult burden to satisfy than that imposed by NAFTA. The Uruguay Round applies the cost-effectiveness standard to all regulations, but NAFTA applies it only to protection of animals and plants. The Uruguay Round does not permit measures that cannot be "maintained against available scientific evidence," whereas NAFTA requires only a "scientific" basis for regulations, so that even if research is ambiguous or mixed, measures based on that research will be upheld. Under GATT, standards "shall not be more trade restrictive than necessary to fulfill a legitimate objective;" this rule permits the balancing of commercial and ecological factors.[42] NAFTA incorporates GATT's obligations by reference.[43]

Under NAFTA, parties can establish levels of protection of human health and ecology based on "legitimate" objectives.[44] The only trade regulations permitted are those that affect "product characteristics or their related processes and production methods."[45] The Bush administration emphasized that NAFTA "explicitly maintains our right to enforce existing U.S. health, safety, and environmental standards."[46] But NAFTA does not expressly provide for that right: Environmental laws can be challenged as discriminatory and can be reviewed by dispute panels. The Bush administration also argued that the agreement "allows the parties, including states and cities, to enact environmental or health standards that are tougher than national or international norms."[47]

One problem is that states may not be able to impose more stringent environmental standards than the federal government, because only the participating nations themselves can impose more stringent standards. If, for example, California had more stringent pesticide residue standards than the rest of the nation, Mexican exporters of food who failed to meet the California standards could challenge them as "arbitrary or unjustifiable distinctions." California itself could not appear before the NAFTA but would have to be represented by the federal government. Mexico could argue that the California standard is a "disguised restriction" on trade, aimed at protecting domestic producers, since no other state has felt compelled to provide that level of safety.[48] Another NAFTA provision, concerning energy, might conflict with environmental protection efforts. Parties are free to "allow existing or future incentives for oil and gas exploration, development, and related activities."[49] But such provisions could be used to restrict efforts to reduce fossil fuel combustion in response to the threat of global climate change.[50]

Nevertheless, in some important ways, NAFTA appears to accommodate environmental values more than other trade agreements. In the dispute settlement process, for example, a party that argues that an environmental or health standard violates NAFTA has the burden of proof, whereas under GATT, the party need only show that there is an inconsistency, and the accused nation must defend its practice. In most cases, under NAFTA, the party whose standards are challenged can choose the forum for resolving the dispute. However, NAFTA does not require that at least some dispute panel members have expertise in environmental matters, nor are panel meetings required to be public.[51]

Opponents of NAFTA fear that stimulating the Mexican economy will increase pollution levels and resource damage; proponents believe that a healthier economy will translate into greater environmental protection. One of the most important provisions of NAFTA is that environmental standards should be "harmonized upward,"

Interaction of Environmental Protection and Economic Growth

that parties should use "international standards . . . without reducing the level of protection of human, animal, or plant life or health."[52] NAFTA requires that a party "should not waive or otherwise derogate" environmental and health standards in order to encourage investors."[53] But the language is the hortatory "should" rather than the obligatory "shall." And trade agreements have been criticized for encouraging standards that satisfy the lowest common denominator.[54]

In sum, the environmental provisions in NAFTA appear to be significantly stronger than those proposed in the current round of GATT negotiations: Parties are freer to choose their own appropriate level of protection; the least-trade-restrictive test is eliminated; harmonization of standards is aimed upward; the three major international environmental treaties are deferred; and the language discourages the relaxing of standards in order to entice investors. Despite these strengths, NAFTA does not list explicit environmental goals; it fails to recognize subnational governments; many provisions are vague; parties are not obligated to comply with environmental treaties; and no mechanism exists to develop regional environmental agreements or to establish common standards for protecting shared air and water. Both supporters and detractors of the treaty can claim support for their side, but NAFTA is at least a good argument for the position that free trade and environmental quality can be pursued together.

The bill that was introduced in Congress to implement NAFTA in the United States did not explicitly provide for environmental safeguards.[55] The NAFTA package includes the agreement itself and three side agreements. The agreement itself permits each country to establish environmental standards that are more stringent than international ones, as long as they are supported by scientific evidence, and to prohibit the import of products that don't meet its standards. The supplemental agreements outline a plan for cleaning up pollution along the U.S.–Mexican border and provide remedies should one country fail to enforce its labor or environmental laws as a way to encourage foreign investment.

BEYOND GATT AND NAFTA

Can free trade and environmental preservation be made more compatible? One option is to include in international trade agreements prohibitions against externalizing specific environmental or social costs. Another approach is to create trading organizations that expressly include environmental and social costs in transactions. The European Union, for example, already has in place some of the regulatory infrastructure that could require full-cost pricing. NAFTA, more than GATT, recognizes environmental concerns; future regional agreements may deal more effectively with environmental concerns as they build on the NAFTA experience to create more opportunities for including environmental considerations in trade.

Balancing future trade agreements with environmental ones will bring difficult trade-offs. Trade agreements have reduced many tariffs and nontariff barriers, but domestic environmental concerns of importance to some nations have been challenged by others as restraints of trade, as demonstrated by the recent conflict between the United States and Mexico over tuna harvesting. We will need to address some hard questions: Should tariffs be imposed on products from countries that fail to meet minimum environmental standards? Can strict domestic environmental regulations be viewed as restraints on trade? Should environmental regulation be subject to trade agreements, or vice versa?

Rethinking Technology

One of the most enduring criticisms of environmental regulation is that it often merely shifts pollution from one medium to another rather than actually reducing pollution(for example, some pollutants are removed from the air but are then deposited as solid waste, which then leaches into groundwater). Environmental regulation in the United States and many other countries has been on controlling or reducing, rather than eliminating, pollution. Many view that approach as the least obtrusive and disruptive to industry practices, the least expensive to comply with, and essential in balancing environmental protection with economic development.

Reduction or control of emissions, however, has a number of weaknesses. Pollution control equipment often breaks down. Control technologies are often expensive to install and maintain. Regulatory agencies must monitor emissions to ensure they fall within accepted guidelines. The debate over acceptable levels of risk is difficult and contentious, and delays implementation of environmental laws and regulations. Virtually every major health-based standard in the United States has been challenged as insufficient to protect human health. Because of the sparse distribution of air-quality monitors, many people may be exposed to much greater levels of pollutants than standards recommend.[56]

One of the first environmental laws enacted in the United States, the National Environmental Protection Act establishes as national policy the encouragement of efforts "which will prevent or eliminate damage to the environment."[57] But that act is much better known for its procedural mandates than for directly requiring environmental protection.

TECHNOLOGY CHANGES AND POLLUTION CONTROL

Rather than trying to control emissions, a much more effective way to protect human health and ecological systems is to switch to technological processes and materials that eliminate pollution. Such an approach requires the availability of appropriate technologies, economic resources to make the required changes, and political will to overcome resistance to change. Barry Commoner argues that the only real successes have come when basic changes in technology and production have occurred. "Every pollutant on the very short list of real improvements—airborne lead, DDT and related pesticides, PCB, mercury in the Great Lakes' fish, and Strontium 90," Commoner argued, is a result of changes in production: "In each case, the production process that originally generated the pollutant has been changed."[58]

Commoner argues that the real problem is that after World War II, the way goods were produced changed: disposable instead of reusable containers, chemical fertilizers and pesticides instead of organic ones, synthetic detergents rather than soaps, synthetic rather than natural fibers, and plastics instead of paper are all examples of the shift in production that became a major assault on the environment. Perhaps most significant has been the change to more powerful motor vehicle engines that combust fuel at higher temperatures, thus producing nitrogen oxides and requiring fuel additives that produce toxic air pollutants.[59]

Fundamental changes that go well beyond new technologies to reach economic and social institutions are required: Solving the environmental crisis is "fundamentally a political problem because it calls for the establishment of a new, social form of governance over decisions that are now exclusively in private, corporate hands."[60] Principles of ecology alone cannot generate agreement over a new social consensus. The crisis is rooted in the technology of production. It is there that we must find remedies, rather than believing that limiting population growth, organizing society along ecologically sound principles, or living in self-sufficient bioregions will solve these problems.[61]

For Commoner, the solution is the "proper choice of technologies and systems of production" that will redirect industry from short-term profits to long-term concerns of environmental quality.[62] Commoner and others argue that we must view corporations primarily as social institutions because of their impact on local communities. Communities must be able to demand that corporations benefit society as well as stockholders.[63] A similar problem has occurred in the former communist countries of Eastern Europe, where the same, albeit older, technologies were used and short-term profit maximization was the overriding goal. Despite a collectivist ideology, at least in theory, the public was essentially not involved in decisions as to which technologies were to be used and how pollution was to be prevented or remedied.[64]

Markets are effective in facilitating the distribution of goods and in choosing the products that best suit our needs. But markets have not dealt effectively with the social consequences of the production modes employed in most industries. Prices should not depend on profit maximization alone but must reflect a broader range of concerns such as ecological sustainability. For Commoner, the industrialized nations need to move quickly to encourage an open

Interaction of Environmental Protection and Economic Growth

discussion of the "serious conflict between our unexamined capitalist ideology and the failed effort to resolve the environmental crisis—as a prelude to radical (in the sense of getting at the root of the problem) remedial action."[65] Principles of corporate responsibility (such as those developed after the *Exxon Valdez* oil spill in Alaska) are one way to ensure that community concerns affect corporate decision making. Given their history of opposition to regulation, industries will not, of course, accede to the loss of power easily. However, in the future much more will be asked of them than current regulatory programs demand. Even if one believes that immediate action is not required to protect the biosphere, increased conservation efforts are inevitable.

The technological basis for transforming the present systems of production to ecologically sound ones is already available, Commoner argues. The transition will take time, but changes in the current system can reduce environmental damage and pave the way for greater changes. Natural gas, for example, produces less sulfur dioxide and carbon dioxide emissions than do coal or oil. Transition efforts, according to Commoner, ought to have three general purposes: (1) prevent local pollution, (2) prevent worldwide environmental effects (such as global warming), and (3) accelerate ecologically sensitive development in the poorer nations. Efforts that address only one of these goals are problematic: Increased reliance on nuclear power, for example, would reduce greenhouse gas emissions but would increase the risk of catastrophic releases of radioactive material and the amount of radioactive waste to be disposed of. Conservation of fossil fuels, although essential, may hamper economic development in the developing world.[66]

The availability of feasible substitutes for the products and processes that cause global environmental problems will also affect regulatory efforts. Generally, the greater the availability of adequate substitutes, the more likely that industry opposition can be overcome, economic disruptions minimized and regulatory restrictions put in place. Industry scientists, for example, have been quite optimistic about developing alternatives to CFCs.[67] Some alternatives to fossil fuels do not contribute to global warming. Toxic wastes can be treated and stored without leakage. Several strategies for improving energy efficiency are widely accessible and applicable. Many of the barriers are economic rather than technological; these alternatives are sometimes more expensive than traditional practices, or at least impose high initial costs that take some time to recoup.

THE POLITICAL CONTROL OF TECHNOLOGY

Because they are so interdependent and so independent of political boundaries, environmental problems and solutions pose tremendous challenges for international policy making and national sovereignty. Economic considerations are obviously critical in determining whether regulatory initiatives will be pursued. Even when the benefits of regulation appear to outweigh the costs, the distribution of benefits and costs within national economies will produce some winners and some losers. Any redistribution of these burdens will produce corresponding political pressures.

Democratic governments are arising throughout the world, and some of their energy and optimism may spill over to the industrialized democracies, which have largely failed to make their industries susceptible to democratic control. The debate over environmental regulation involves clashing ideologies and issues; however, this kind of debate rarely takes place in democratic elections. The governance of production, a fundamental issue in society, unites concerns about environmental quality with other major social problems—poverty, discrimination, inequality, opportunity, political openness and competition.[68] E. F. Schumacher's *Small Is Beautiful* has become a classic statement on the environmentally "destructive forces" that are unleashed by the "logic" of modern industrial production and that can only be brought under control if production technologies are transformed.[69]

Environmental problems and solutions transcend government powers. Corporate executives make most of the decisions that affect resource consumption and pollution production. The social impact of business decisions overwhelms decisions made by governments. In fact, many business decisions generate the need for public policies—governments enact policies to mitigate the effects of business decisions made in pursuit of corporate profits. However, corporate executives regularly translate their tremendous economic resources into political power. Elected officials are so dependent on public perceptions of the economy that they face tremendous pressure to promote the interests of polluting industries. Businesses have considerable resources to help them resist public demands to alter their operations. Threats of job losses and plant shutdowns are very powerful, and elected officials are quick to respond. In a political contest between the immediate danger of job losses and a (somewhat uncertain) warning of future environmental crisis, the former usually wins. Indeed, Barry Commoner argues that environmental problems cannot be understood apart from issues of the social governance of production. Industry efforts to reduce production costs and maximize short-term profits by exporting pollution threaten public health and ecosystems.[70]

Since corporations are the dominant economic institution, their commitment to improving environmental quality is critical. They are already creatures of government, as state and federal laws govern their creation and operation. Early corporate charters were restrictive; by contrast, contemporary corporate charters provide few restrictions. Paul Hawken argues that it is not enough to try to regulate their behavior; we need to change the basic rules of commerce. Corporations lobby to weaken environmental laws because compliance with those laws will cost shareholders profits. Corporations, unlike private citizens and nonprofit groups, can deduct lobbying expenses from their taxable profits. Corporations should be subject to regulation because they are created by government and given important legal and economic benefits. If they repeatedly disobey the law, they should be disbanded and their assets sold and given to stockholders after the damage they have caused is remedied.[71]

Environmentalists concerned about damage to ecosystems, champions of workers who are exploited by industries, community activists who find that their communities are at the mercy of external economic forces, social and religious leaders who fear the consequences of materialism, peace advocates who protest the export of arms rather than assistance, representatives of indigenous peoples whose lives and livelihood are threatened by the unbridled harvesting of natural resources, and proponents of true democracy all join in calling for a fundamental rethinking of our traditional deference to large corporations. Governmental power needs to be sufficiently strong to allow people to make basic choices about what kind of community they wish to live in and how they can protect the natural, social, and cultural preconditions for the kind of society they seek. Government must ensure that communities can protect human health and the environment. Communities need to be empowered to ensure that markets require total internalization of costs and that companies compete to be more socially and economically responsible, so that industrial activity is consistent with our natural, ecological means.[72] This does not require a change in human nature: Restoring the environment can become an essential part of what industry does, and markets can be forced to recognize the true costs of producing goods.[73]

Commoner criticizes the leaders of mainstream environmental groups for failing to recognize the need for fundamental change and for being too quick to accept the political status quo. These groups have increasingly taken a pragmatic position in working with industry interests, emphasizing political accommodation, participating in industry-environmental group roundtables and meetings, and in arguing that environmental "arguments must translate into profits, earnings, productivity, and incentives for industry."[74] In contrast, some grassroots groups have become much more aggressive in demanding that there be no environmental damage in their neighborhoods. Commoner asserts that "the environmental movement is split in two":

> The older national environmental organizations, in their Washington offices, have taken the soft path of negotiation, compromising with the corporations about how much pollution is acceptable, and sometimes helping to market their products, even when they are ecologically unsound. The people living in the polluted communities have taken the hard path of confrontation, demanding not that the dumping of hazardous waste be slowed down, but stopped; not that dioxin-producing

incinerators be equipped with unworkable emission controls, but abandoned in favor of recycling. The national organizations deal with the environmental disease by negotiating about the kind of "Band-Aid" to apply to it; the community groups deal with the disease by trying to prevent it.[75]

Commoner argues further that our energies are consumed by issuing standards for controlling pollution, justifying those standards, and then defending them in court when they are (inevitably) challenged. All this activity has had "little impact on environmental quality." In neighborhoods throughout the nation, in contrast, citizens confront polluters:

> Banded together in an ad hoc committee under an acronym such as STOP, RAGE, or WASTE, the local residents fight not for improved standards of exposure but for no exposure at all. Their preferred standard of exposure is zero. They want pollution prevented rather than controlled.[76]

To make such changes we need new ways to get citizens involved in decisions about the consequences of technology. Decisions about technologies are inherently political ones, since they have such profound impacts on social life.

Paul Hawken argues that we need an open process of education, dialogue, redesign, and shared decisions, since the primary global environmental challenge is political transformation. We need to create new deliberative processes that bring everyone together—environmentalists, the poor, people threatened with losing their jobs, victims of pollution, and industry officials. Governing needs to be broadened democratically, but participation of industry lobbyists in regulatory proceedings is not democracy.[77]

The Prospects for Eco-nomics

Concerns about the competitiveness of business have played a major role in the debate over environmental regulation. The economic costs of complying with regulation are politically important, for several reasons. They are disruptive to and threaten established economic interests. They are a major burden on old, dirty industries, especially those that have delayed modernization. They will undoubtedly produce layoffs, job loss, and other problems because some sources of pollution must be eliminated to protect public health. Companies that are having trouble competing in global markets will find environmental laws and regulations an easy excuse, when the real problem may have been poor management or other difficulties.

Environmental regulations are often not enforced because of the economic disruptions they cause. However, if we can find ways of getting beyond the short-term disruptions, we will find that economic and environmental objectives can for the most part be compatible. Pollution is waste: The less pollution produced, the fewer the resources used. Companies that are more efficient and that use less energy will be more competitive in the long run. Global and domestic demand for cleaner processes and products can only increase, given trends in pollution and environmental degradation, and clean companies will have greater opportunities. It is crucial to recognize that regulation shakes things up, threatens those who are slow to modernize and who have pursued short-term profits over long-term profitability and sustainability. Costs can be put off, problems can be buried or dispersed in tall smokestacks, but only temporarily. Investments that prevent or reduce pollution will eventually mean less waste, lower cleanup costs, reduced legal liability, and other benefits. One of the great challenges for policy makers is to provide the sanctions and incentives for modernization and improved efficiency, and to create regulatory programs that serve as a catalyst for change.

An even more powerful argument may be that incentives and sanctions that force U.S. firms to become more efficient and less polluting or wasteful may have great long-term economic benefits. Other countries are already moving in this direction. Japanese and German companies are major players in pollution-control equipment markets, but U.S. firms have an opportunity to make real progress in global markets. As resources become more scarce,

The Prospects for Eco-nomics

prices will increase, and more efficient, less wasteful processes will translate into lower prices. American pollution-control companies sold $130 billion worth of goods in 1991. Worldwide sales reached $370 billion. The United States has a positive trade balance with all its major trading partners for these kinds of products, and the growing awareness of environmental degradation in Eastern Europe and elsewhere will stimulate even more demand for pollution-control devices and processes.[78]

Industries can choose not to accept the traditional balancing of economic activity with inevitable pollution, but to find ways of pursuing environmental and economic goals together. However, the economic benefits will not be distributed as are existing benefits. This is a major challenge to political leadership. Environmental protection need not be synonymous with economic decline but can be consistent with a dynamic, competitive economy. Of course, some wasteful practices and approaches will have to be abandoned, and some will be at least temporarily displaced. Governments can institute appropriate policies to facilitate this transition, and in so doing, reduce the opposition to change. Despite these and a host of other challenges, a number of policy developments will likely play important roles in the implementation of international environmental agreements in ways that balance economic and environmental goals.

The Rise of Nonstate Institutions

One of the most important global trends is the rise of nonstate actors such as multinational corporations, nongovernmental institutions, and international organizations. The new global system includes a host of actors in addition to the traditional nation-states. Nations need to devise new partnerships with industry, with scientists, and others. One example is the Intergovernmental Panel on Climate Control, in which many of the world's leading scientists work on behalf of the UN and participating states, but are independent of any governmental direction. Other innovations such as small-scale banks for microenterprises and investments in energy efficiency and the development of new technologies are also essential.

Another example is government-business cooperative efforts to reduce pollution, where government agencies provide technical assistance rather than emphasizing enforcement. Twenty years ago, the 3M Company developed its Pollution Prevention Pays program, an integrated effort to redesign manufacturing to reduce emissions. By 1990, the company had saved $537 million and had reduced air pollution by 120,000 tons, waste water by 1 billion gallons, and solid waste by 410,000 tons. In 1986, the plan was amended to include the goal of eliminating all emissions by the 1990s. Other companies are integrating processes: for example, waste heat in the form of steam is being recycled into other facilities. Progress has been made in the manufacture of appliances, vehicles, and other areas. Monsanto has promised to cut its emissions by 90 percent.[79] Much more can be done to combine industrial innovations with public policies that create strong incentives to continue and expand these efforts, and to see that this occurs internationally.

Reuse of Products

Other changes could revolutionize industry. Instead of buying products, for example, consumers could simply lease them. Appliances, motor vehicles, and other products would not be disposed of but would be returned to the manufacturer for dismantling, reusing, or reclaiming of materials. This would carry us far beyond recycling to a fundamental redesign of products that would take into account their eventual return and reuse by the manufacturer. Labor costs would increase, and material and energy use would decrease, exactly what we need in a world of high unemployment and scarce resources.

Wastes that cannot be reused could be stored under tightly controlled conditions in public facilities, where producers would rent storage space. The cost of renting space would provide a clear incentive for companies to reduce their production of toxic materials, and controlled facilities decrease the chance that the substances will con-

The Prospects for Eco-nomics

taminate land and water. These "radical" changes focus on the root causes of pollution and provide incentives for innovative solutions that protect the environment, create jobs, and reduce costs.[80]

True-cost markets will also stimulate local production and economic community as the real costs of transportation make long-distance shipping more expensive. Local companies will become more competitive, since they are more likely to be managed by local residents, local environmental conditions would become more important in corporate decision making. Small companies are more likely to be environmentally friendly, since mass markets and mass production cannot easily be shaped to reflect natural conditions. Small companies are more likely to be accountable to local citizens: The closer we are to a production site, the more likely we are to know how production occurs and the more likely we are to take responsibility for it.[81]

Tax Policy

Tax policy creates incentives and disincentives for a host of activities that affect economies and environments. Taxes in industrial countries on payrolls, investment, and other productive activities are increasingly being criticized as making economic problems worse. In the United States, for example, companies can deduct the cost of investments in new equipment, but individuals cannot deduct their expenses in improving their education; as a result, automation is encouraged and education is discouraged. Taxes on sales, income, payrolls, and profits create a disincentive for these activities. It makes little economic sense to tax activities that are socially beneficial.

The environmental consequences of tax policy are equally unfortunate. Governments "tend to undertax destructive activities, such as pollution and resource depletions and environmental quality."[82] Policies that permit industries to dump their waste on other people encourage pollution production. This also imposes economic costs on others in the form of reduced property values, cleanup costs, and medical expenses, not to mention the adverse impact on health itself. The failure to tax air and water pollution ensures that polluters will avoid the full costs of production.

Just as perverse are public policies that subsidize environmentally destructive activities rather than taxing them. Logging, mining, pesticide and fertilizer use, and energy development are subsidized rather than taxed, encouraging inefficient production and consumption. According to one estimate, countries provide some $800 billion a year in subsidies for environmentally damaging activities.[83] If taxes were imposed on excess profits from natural resources, for example, producers would still earn profits and have an incentive to produce, but society would gain the windfall as resources become increasingly scarce, forcing prices up.

Tax policy can be a clear means of integrating and improving economic and environmental policy. Shifting taxes away from desirable actions such as earning profits and paying salaries, and directing them toward undesirable actions such as producing pollution and harvesting scarce resources, can strengthen economies and make them more ecologically sustainable at the same time. Several kinds of policies are critical. Exploitation of resources that generate windfall profits should be taxed, instead of income and sales. Wood from virgin forests, ocean fish, and other resources are become increasingly scarce, whereas harvesting costs have often fallen due to economies of scale and new technologies. Profits often exceed fair rates of return, and those excess profits should be taxed and shared with the entire society. Political leaders of many countries use access to public resources as political payoffs, through leases to mining and timber companies, for example, rather than as a way to raise needed revenue.

Pollution and depletion of scarce resources should be taxed. Taxing pollution has several advantages: It ensures that producers of pollution take some responsibility for the harm they cause; it guarantees that those who benefit economically from industrial production also pay the costs and do not pass them on to others who do not enjoy the benefits; it creates clear incentives to reduce harmful activities without the heavy hand of government regulators and the inherent loss of flexibility and freedom that comes from

The Prospects for Eco-nomics

command and control regulation, and it encourages pollution reduction to be efficient. Governments can impose taxes on greenhouse gas emissions, fish catches, the building of homes on habitats of endangered species or other actions that harm ecosystems and threaten biological diversity and stability, and timber sales in virgin forests. Taxes can permit us to include in current price calculations the interests of future generations who, if present, would also demand clean air and water and old-growth forests and would force up the prices for them. It is difficult to calculate the appropriate level of these taxes, since pollution levels differ significantly across similar sources (motor vehicles vary tremendously in their emissions, for instance, as a result of kinds of fuel, meteorological conditions, driving patterns, and other factors). Placing economic values on environmental quality or scarce resources is similarly difficult. Taxes must be integrated with other laws and policies. The benefits of increased gasoline taxes, for example, are countered by land-use decisions that encourage urban sprawl and more driving. Increasing taxes may not solve the problems of how pollution sources are distributed and the tendency to concentrate them in low-income communities. Some endangered species will require absolute protection and a ban on their killing.[84]

International agreements are needed to ensure that industrial activities take place under common rules, especially tax policy. The most promising agreements are those that include decentralized implementation and enforcement mechanisms based on true cost pricing, rather than centralized bureaucratic schemes. Although the cost to consumers increases, ensuring that prices reflect the true costs of production is essential. As savings result from reduced waste, however, prices to consumers will fall.

Ending environmentally damaging subsidies, increasing taxes on pollution, cutting taxes on payroll, and other desirable initiatives are championed as ways to stimulate economic growth and pursue environmental and social goals. Some studies argue that from $500 billion to $1 trillion in resources could be made available by ending subsidies on energy, pesticides, and fertilizers; resource development; and other actions that contribute to environmental harm.

But it is difficult to make sure such changes achieve their goals: A cutback in subsidies for kerosene use in Ghana, for example, led to more cutting of trees in mangrove forests that are critical habitats for marine life and stabilize coastlines. Agricultural subsidies cost $335 billion in the industrialized countries, according to one estimate, but ending them would threaten small farmers. Some subsidies encourage energy conservation and the development of renewable energy resources, but distinguishing between good and bad subsidies, and anticipating the consequences of ending them, is an extremely difficult policy task.[85]

Emission taxes on pollution and fees, if all countries agree to them, can make important contributions to environmental quality. Sweden has levied fees on coal, oil, and peat used for fuel in order to raise money and encourage conservation. Mauritania has increased fees for fishing in its 200-mile economic zone. Cuba imposes a tax on sugar plants that exceed allocated energy use. An energy company in Colombia makes annual payments to a *campesino* cooperative to encourage them not to cut down trees, thereby contributing to soil erosion and sedimentation that would threaten the operation of the cooperative's hydroelectric plant. Costa Rica charges a 15 percent tax on gas and fossil fuels that generates money for reforestation efforts. The main impetus for these kinds of reforms come from NGOs. Governments of developing countries as one economist put it, "are as disposed to protect the green environment as the currently developed countries were fifty or sixty years ago." The lack of support for green taxes by industry is compounded by their limited applicability: The taxes can be used to limit exploitation, but not actually to protect resources against development. They also have distributional impacts, and those who are likely to end up losing from the reforms have a strong incentive to block them. A proposed international carbon tax on fossil fuels could generate funds for sustainable development projects and create an incentive for conservation, but political opposition to such a tax is daunting.[86]

Some argue that green taxes should be revenue neutral; they should not be used to reduce budget deficits, for example, because that will make them unpopular. Other taxes could be decreased to match the green tax increases. The main goal of green taxes should be to ensure that prices communicate accurate information about the costs of production. For example, taxes can be levied on the use of virgin materials. Markets do not distinguish between sustainably harvested lumber and that which results from unsustainable clear-cutting. Other taxes could be levied on emissions, environmental damage, and consumption of fossil fuels and other nonrenewable resources. These taxes could be adopted incrementally, over a period of twenty years or so, in order to provide time to make adjustments. They would create incentives to improve the efficiency of production and distribution. They would also permit consumers to reassess their decisions and to adapt their behavior to environmental constraints.

Those who produce natural resources should increase their prices to reflect true costs, but government policy is necessary to ensure everyone does this; otherwise those who externalize will continue to benefit. Current green taxes, because they are so low, are used mostly to raise resources for environmental programs, rather than as an alternative to regulation. Industries oppose green taxes because they will eventually force companies to change the way they operate.[87] Our obsession with the cheapest price for everything, which permits firms to externalize many of their costs, has actually dampened innovation in some cases. More efficient companies are not necessarily rewarded in the market system. In contrast, every transaction should provide constructive feedback on resource use.[88]

Shifting costs to the marketplace makes them more visible and enables us to decide how to respond. Markets that are based on true-cost prices can help us choose options that give us the greatest future choices.[89] For example, if we wait too long, it may will be too late to prevent climate change; if we convert now to a solar economy and later find out there was no threat, we will still be better off.[90] Taxing energy would produce immediate as well as long-term environmental and economic benefits. Coal may be our most expensive form of energy when we consider the threat of global warming, acid rain, water pollution, and black lung disease, so the green tax on it would be relatively high. Economic and environmental concerns can be compatible, since conservation reduces costs: If the United States were as energy efficient as Japan, we would have spent $200 billion a year less in energy costs during the past decade.[91] And energy efficiency creates more jobs than building new power plants. Similarly, we can tax hyrodcarbon-based chemicals, and encourage their replacement with processes using organic, nonpolluting, renewable resources. Farmers who use sustainable approaches in agriculture would be able to make a profit if we ensured their competitors could not externalize their costs through the heavy use of pesticides and fertilizers, which threaten land and water quality.[92]

Future Costs and Benefits

Making markets work through true cost pricing is critical, but it may not be enough. Markets have several important flaws. They understate or discount the value of costs and benefits that extend very far into the future. As a result, "critical ecological resources that will be essential for our well-being even 30 years from now not only have no value to rational economic decision makers, but scarcely enter their calculations at all."[93] Markets don't account for real scarcity:

> Although the market price mechanism handles incremental change with relative ease, it tends to break down when confronted with absolute scarcity or even marked discrepancies between supply and demand. In such situations (for example, in famines), the market collapses or degenerates into uncontrolled inflation, because the increased price is incapable of calling forth an equivalent increase in supply.[94]

Markets also fail to respond to the problems of ecological scarcity because scarcity "tends to induce competitive bidding and preemptive buying, which lead to price fluctuations, market disruption, and the inequitable or inappropriate distribution of resources." Consumers may not respond to rising prices in ways predicted by economic theory. Some consumer decisions are rather independent of price increases. For example, increasing the price of gasoline is likely to have little long-run impact on driving, unless price increases are so dramatic that they discourage people from driving their own vehicles. If such price increases occurred, the resultant dilemmas of social equity and resource allocation would be daunting. Other consumer decisions, such as choosing the kind of energy to heat a home, are essentially locked in because of the high capital costs involved in converting to another energy system.

High prices may not be enough to deter ecologically unsustainable activities.[95] In the 1980s, for example, commodity prices fell, despite growth in population and demand. However, this does not mean that scarcity is not a problem. Prices do not tell us the true costs of our consumption. The rate of harvesting of natural resources increased tremendously as competition expanded and as countries with natural resources sought earnings to reduce debt. We are depleting resources rapidly, and current prices do not reflect scarcity in the future. Inventories of natural resources show that depletion is occurring on such a scale that shortages are inevitable. We are borrowing from the future to sustain our current consumption.

Economists believe that most things can be substituted for each other. Markets reduce all items to a dollar value so they become interchangeable. But the idea of ever-expanding abundance is based on economic theory, not on science or biology or nature. Daly and Cobb argue that we need to shift to an economics based on its root word, *oikonomia*, which means the management of the household, in order to increase its value over the long run and to all members. Economics for community expands this notion to include the larger community of land, resources, biomes, languages and cultures, and institutions.[96]

Traditional economic concepts are not sufficient for an ecologically sustainable economy:

> Economists assume the future will be much like the past. Since new markets and technology have avoided catastrophe in the past, we can count on them to do the same in the future. Ecologists believe they see unique problems in the future, which will demand solutions outside the capacity of our present market

mechanisms. Economists tend to see evolution as a series of continuous reversals: problems leading to solutions, new problems leading to new solutions. Ecologists are worried about irreversibilities. When species are lost, no change in price or technology will bring them back.[97]

Budget and trade deficits get much more attention than the global environmental deficit, the "collective and mostly unanticipated impact of humankind's alteration of the earth's atmosphere, water, soil, biota, ecological systems, and landscapes." This deficit has occurred because "the longer-term ecological, social, and economic costs to human welfare are greater than the shorter-term benefits flowing from these alterations."[98] Like budget deficits, environmental deficits rob future generations as they permit profligate consumption by the current generation. The solution is to make clear and compelling the link between ecological and human welfare. We can no longer assess economic activity from the perspective of short-term returns, but must constantly view it through long-term lenses.[99]

One preliminary but promising such effort is the World Bank's proposal to measure a country's wealth by estimating not only how much it produces, but also its investments in natural and human resources. The new system breaks down national wealth into three major attributes:

> "natural capital," or the economic value of timber, mineral deposits, land, water and other environmental assets;
>
> "produced capital," or the value of a nation's machinery, factories, roads; and
>
> "human resources," such as the educational level of a population.

National wealth is ultimately viewed as the value of produced goods minus consumption, depreciation of produced assets, and use of natural resources.

Several advantages arise from this kind of broad balancing of national economic books. First, it focuses attention on the value of investments that increase human capital. Education, health care, and other social services are critical in development human resources. Second, as data unfold over time, relative changes in a country's development can be identified and adjustments made. Third, gathering information on natural resource wealth can help identify the long-term consequences of selling off natural resources in the short term: Although rapid harvesting of resources might appear as gains in immediate economic figures such as gross national product, it will be considered a reduction in indicators of natural resource wealth.[100]

In most issues of economics and environment, the real questions are moral ones. We have been willing to accept a certain level of pollution, together with the resultant health and environmental harms, in return for material progress. We seem resigned simply to hope that the benefits will outweigh the harms. We have ignored risks to current and future generations because of the sacrifices that remedies may require. We cannot escape difficult choices, but we have great possibilities for finding effective, creative ways to pursue effective economic and environmental policies. The growing debate over the need for a new political economy paradigm places the health of the biosphere as the central concern. If such a dramatic change in economics and politics is possible, we must reconcile an agenda for radical reformation with the incrementalist approach, which seems to be more attuned to our political, economic, and social possibilities.

References

1. John Noble Wilford, "What Doomed the Maya? Maybe Warfare Run Amok," *New York Times*, 19 November 1991; David Roberts, "The Decipherment of Ancient Maya," *Atlantic Monthly* (September 1991): 87–100.
2. Julian Burger, *The Gaia Atlas of First Peoples* (New York: Anchor, 1990), 15–17.
3. For a discussion of the quote's authenticity, see David Suzuki and Peter Knudtson, *Wisdom of the Elders* (New York: Bantam, 1992), xx–xxiii.
4. T. C. McLuhan, *Touch the Earth* (New York: Simon and Schuster, 1971), 1.

5. Ibid., 53.
6. Ibid., 22.
7. Ibid., 107.
8. See, for example, Mary E. Clark, *Ariadne's Thread* (New York: St. Martin's Press, 1989): Carolyn Merchant, *Radical Ecology* (New York: Routledge, 1992); and Bill McKibben, *The End of Nature* (New York: Random House, 1989).
9. See, for example, Gregg Easterbook, *A Moment on Earth* (New York: Viking, 1995); Paul Hawken, *The Ecology of Commerce* (New York: Harper Business, 1993); and Jim Macneill, Pieter Winsemius and Taizo Yakushiji, *Beyond Interdependence: The Meshing of the World's Economy and the Earth's Ecology* (New York: Oxford University Press, 1991).
10. This debate is effectively addressed in Walter A. Rosenbaum, *Environmental Politics and Policy* (Washington, D.C.: Congressional Quarterly Press, 1993); and Michael E. Kraft, *Environmental Politics and Policy* (New York: HarperCollins, 1995).
11. See Theodore Panayotou, *Green Markets: The Economics of Sustainable Development* (San Francisco: ICS Press, 1993).
12. See, for example, Wallace Kaufman, *No Turning Back: Dismantling the Fantasies of Environmental Thinking* (New York: Basic Books, 1994).
13. See Robert Repetto, *Jobs, Competitiveness, and Environmental Regulation: What are the Real Issues?* (Washington, D.C.: World Resources Institute, 1995).
14. Ronald Bailey, ed., *The True State of the Planet* (New York: Free Press, 1995).
15. See Aaron Wildavsky, *But Is It True? A Citizen's Guide to Environmental Health and Safety Issues* (Cambridge, Mass.: Harvard University Press, 1995).
16. Hawken, *The Ecology of Commerce*, 97–99.
17. Ibid., 197.
18. For more on trade and environmental issues, see Daniel C. Esty, *Greening the GATT: Trade, Environment, and the Future* (Washington, D.C.: Institute for International Economics, 1994); Office of the President, *The NAFTA: Expanding U.S. Exports, Jobs and Growth* (Washington, D.C.: GPO, 1993); Durwood Zaelke, Paul Orbuch, and Robert F. Housman, eds., *Trade and the Environment: Law, Economics, and Policy* (Washington, D.C.: Island Press, 1993); Ralph Nader, et al., *The Case Against Free Trade: GATT, NAFTA, and the Globalization of Corporate Power* (San Francisco: Earth Island Press, 1993); U.S. Congress, Office of Technology Assessment, *Trade and the Environment: Conflicts and Opportunities* (Washington, D.C.: GPO, 1992); Environmental Protection Agency, *The Greening of World Trade* (Washington, D.C.: EPA, 1993).

References

19. Robert Costanza et al., "Sustainable Trade: A New Paradigm for World Welfare," *Environment* 37 (June 1995): 16–20, 39–44, at 18.
20. Costanza, "Sustainable Trade," 20, 39.
21. *General Agreement on Tariffs and Trade*, Article III.
22. Ibid., Article XX(b),(g),(d).
23. Steve Charnovitz, "Dolphins and Tuna: An Analysis of the Second GATT Panel Report," *Environmental Law Reporter* 24 (October 1994): 10567–87, at 10583.
24. 16 U.S.C. sec. 1361–1421h, at sec. 1371.
25. 16 U.S.C. sec. 1371.
26. 16 U.S.C. 1371(a)(2)(B)(II), 1371(1)(2)(C).
27. 16 U.S.C. 1415.
28. Charnovitz, "Dolphins and Tuna," 10570.
29. Ibid., 10575.
30. Ibid., 10576–77.
31. Ibid., 10581–82.
32. Ibid., 10585, note 264.
33. Despite the problems, MMPA seems to have been a success. Dolphin mortality in the eastern tropical Pacific has dropped dramatically; less that one dolphin, on average, is killed at each tuna net dropping, and it may be impossible to avoid some accidental dolphin deaths. a strict dolphin policy imposes a cost on other sea life, for example, but neither MMPA nor GATT proceedings provide a forum for exploring such issues. See Charnovitz, "Dolphins and Tuna," 10685–87.
34. David E. Sanger, "World Trade Group Orders U.S. to Alter Clean Air Act," *New York Times*, 17 January 1996, C1.
35. Alan Tonelson and Lori Wallach, "Overruled by the World Trade Organization," *Washington Post National Weekly Edition*, 13–19 May 1996, 23; Bruce Stokes, "Up and Crawling," *National Journal* (30 March 1996): 709–12.
36. David E. Sanger, "U.S. Rejects Role For World Court In Trade Dispute," *New York Times*, 21 February 1997, A1.
37. Peter Passell, "Trade Pacts by Regions: Not the Elixir As Advertised," *New York Times*, 4 February 1997, C1; Calvin Sims, "Chile Will Enter a Big South American Free-Trade Bloc," *New York Times*, 26 June 1996, C2.
38. *North American Free Trade Agreement*, art. 715.3(b).
39. Ibid., art. 715.2.
40. Ibid., arts. 709, 712.1, 712.5. See Steve Charnovitz, "GATT and the Environment: Examining the Issues," *International Environmental Affairs* 4 (Summer 1992): 203, 212–15.
41. NAFTA, art. 712.3.

42. Steve Charnovitz, "NAFTA: An Analysis of Its Environmental Provisions," *Environmental Law Review* 23 (February 1993): 10067, 10068-69.
43. NAFTA, art. 712.3.
44. Ibid., art 904.2.
45. Ibid., art 915.
46. See, for example, Carla Hills, "The Free Trade Pact Is Good for All of Us: Americans, Mexicans, and Canadians," *Roll Call*, 28 September 1992, 50-51.
47. Report of the Administration on the North American Free Trade Agreement and Actions Taken in Fulfillment of the May 1, 1991 Commitments (18 September 1992), 5.
48. Charnovitz, "NAFTA: An Analysis of Its Environmental Provisions," 10070.
49. NAFTA, art. 608.1.
50. Charnovitz, "NAFTA: An Analysis of its Environmental Provisions," 10071.
51. Ibid., at 10070-71.
52. NAFTA, arts. 713.1, 714.1.
53. NAFTA, art. 1114.2.
54. Charnovitz, "NAFTA: An Analysis of its Environmental Provisions," 10071.
55. H.R. 3450, H. Rpt. 103-369.
56. Barry Commoner, *Making Peace with the Planet*, (New York: Pantheon, 1990), 45-46.
57. *National Environmental Policy Act*, 42 U.S.C. 4321, sec. 2.
58. Barry Commoner, "Let's Get Serious about Pollution Prevention," *EPA Journal* (July-August 1989): 15.
59. Commoner, *Making Peace with the Planet*, 46-48.
60. Ibid., 172-73.
61. For further discussion of environment and technology issues, see George Heaton, Robert Repetto, and Rodney Sobin, *Transforming Technology: An Agenda for Environmentally Sustainable Growth in the 21st Century* (Washington, D.C.: World Resources Institute, 1991); George R. Heaton, Jr., Robert Repetto, and Rodney Sobin, *Backs to the Future: U.S. Government Policy Toward Environmentally Critical Technology* (Washington, D.C.: World Resources Institute, 1994); George R. Heaton, Jr., R. Darryl Banks, and Daryl W. Ditz, *Missing Links: Technology and Environmental Improvement in the Industrializing World* (Washington, D.C.: World Resources Institute, 1992); National Science and Technology Council, "Technology for a Sustainable Future" (Washington, D.C.: NSTC, n.d.); U.S. Congress, Office of Technology Assessment, *Industry, Technology, and the Environment:*

References

Competitive Challenges and Business Opportunities (Washington, D.C.: U.S. GPO, 1994).
62. Commoner, *Making Peace with the Planet*, 214.
63. For a classic statement of corporate social control, see Adolph A. Berle and Gardiner Means, *The Modern Corporation and Private Property* (New York: Macmillan, 1932).
64. Commoner, *Making Peace with the Planet*, 220–21.
65. Ibid., 225.
66. Ibid., 198–99.
67. John Holusha, "Ozone Issue: Economics of a Ban," *New York Times*, 11 January 1990.
68. Commoner, *Making Peace with the Planet*, 234–35.
69. E. F. Schumacher, *Small Is Beautiful: Economics as if People Mattered* (New York: Harper & Row, 1973), 295.
70. Charles E. Lindblom and Edward J. Woodhouse, *The Policy-Making Process*, 3d ed. (Englewood Cliffs, N.J.: Prentice-Hall, 1993), 7–9. See also Commoner, *Making Peace With the Planet*, 226–36.
71. Paul Hawken, *The Ecology of Commerce*, 119–20.
72. Ibid., 167.
73. Ibid., 54–58.
74. Commoner, *Making Peace With the Planet*, 177.
75. Ibid., 179.
76. Ibid., 180–81.
77. Hawken, *The Ecology of Commerce*, 219.
78. The figures quoted are from the Department of Commerce and are cited in Michael Silverstein, "Bush's Polluter Protectionism Isn't Pro-Business," *Wall Street Journal*, 28 May 1992, A19.
79. Hawken, *The Ecology of Commerce*, 60–61, 81.
80. Ibid., 69–70.
81. Ibid., 102–03, 147.
82. David Malin Roodman, "Public Money and Human Purpose: The Future of Taxes," *World-Watch* 8 (September–October 1995): 10–19, at 13.
83. Ibid.
84. Taxes can be used to help workers. Pollution taxes are often regressive since they raise the price of energy, transportation, manufactured goods, and other essentials, and take a larger bite out of the total income of poor households than of wealthier ones. They must be combined with wage and income tax cuts aimed at low-income families, rebates for energy taxes, and other adjustments. See Roodman, "Public Money and Human Purpose."
85. Paula DiPerna, "Five Years After the Rio Talkfest: Where Is the Money?" *Earthtimes* (25 January 1997), http://earthtimes.org.

86. Ibid.
87. Hawken, *The Ecology of Commerce*, 172–75.
88. Ibid., 88–90.
89. For a discussion of how economic accounts might reflect environmental quality and natural resources, see Ernst Lutz, ed., *Toward Improved Accounting for the Environment* (Washington, D.C.: World Bank, 1993); Robert Repetto, *Promoting Environmentally Sound Economic Progress: What the North Can Do* (Washington, D.C.: World Resources Institute, 1990); and Robert Repetto, *Wasting Assets: Natural Resources in the National Income Accounts* (Washington, D.C.: World Resources Institute, 1987).
90. Hawken, *The Ecology of Commerce*, 182–85.
91. Ibid., 277.
92. Ibid., 182–83.
93. William Ophuls and A. Stephan Boyan, Jr., *Equality and the Politics of Scarcity Revised* (New York: W. H. Freeman, 1992), 219.
94. Ibid.
95. Ibid., 220.
96. Herman E. Daly and John B. Cobb, Jr., *For the Common Good* (Boston: Beacon Press, 1989), 138.
97. Dennis Meadows, "Biology and the Balance Sheet, *Earthwatch* (July/August 1992).
98. F. Herbert Borman and Stephen R. Kellert, *Ecology, Economics, Ethics: The Broken Circle* (New Haven: Yale University Press, 1991), xii.
99. For a broad overview of how economics might be integrated with ethical, social, and ecological concerns, see Paul Ekins, *The Gaia Atlas of Green Economics* (New York: Anchor Books, 1992).
100. See World Bank, *Monitoring Environmental Progress: A Report on Work in Progress* (Washington, D.C.: World Bank, 1995); Ismail Serageldin, "Third Annual World Bank Conference on Environmentally Sustainable Development" (paper prepared for the Third Annual World Bank Conference on Environmentally Sustainable Development, Washington, D.C., October, 1995).

6

Environmental Preservation and the Less-Developed World

THE INCA EMPIRE, called *Tahuantinsuyu*, or "Land of the Four Quarters," was the largest nation on earth at the time of Columbus's landing in the New World. It included parts of Chile, Argentina, Bolivia, Peru, Ecuador, and Colombia. The empire was divided into eighty political provinces linked by an extensive road system. Dissident villages were relocated to reduce the threat to the kingdom. A long series of wars led to the subjugation of many neighboring regions. The offspring of conquered peoples were brought to Cuzco, the capital, for education and training, then sent back to their homelands. The empire, however, was relatively short lived. Inca civilization collapsed as a result of conflict with neighboring empires, diseases introduced by the Europeans, and political instability. By the time Pizarro and his men arrived in Cuzco, the Incas were already severely weakened, and a band of two hundred or so conquistadors were able to plunder the heart of the empire.

One of the most remarkable aspects of Inca civilization was the great extremes in geography and climate the Incas were able to master: coastal deserts, jungle, and the second-highest mountain range in the world, each requiring different adaptive strategies.

Elevated solar radiation, cold high winds, earthquakes, harsh terrain, poor soils, erratic rainfall, a short growing season, and low oxygen levels all combined to challenge the Incas. The Incas were remarkably skilled at adapting to and flourishing under these challenges, but this knowledge was largely lost with the collapse of the empire. According to one estimate, the Incas farmed from 35 to 85 percent more land than their descendants, the Quechua Indians, do now. Soil erosion has resulted from overgrazing, and much of the land is now poor and infertile. Unlike their ancestors, many of the Quechua live in poverty, no longer in harmony with their land.[1]

Implementing global environmental agreements becomes even more challenging when we consider the special circumstances and concerns of the less-developed countries. These nations are beset by basic economic, social, and political problems that appear to many of their residents and government leaders more urgent than environmental concerns. They are plagued with old, dirty technologies and have fewer resources to make needed changes in industry, agriculture, and conservation than do their wealthier neighbors. Exploding population growth in these countries threatens to overwhelm progress made in conserving resources and reducing pollution. Few of these countries have the regulatory infrastructure—laws, regulatory programs, research capabilities, and enforcement mechanisms—on which to build a new set of requirements for implementing global agreements.

Some residents of these countries believe that they now have the right to consume resources and ignore environmental quality issues much as the developed world did in its path to modernization. They argue that they should be free to burn their large reserves of coal and harvest their tropical rain forests so that they can enjoy some measure of economic growth. But the LDCs also have more to lose than do the industrialized nations. Although the developing countries are different from each other, they generally have few resources to mitigate the effects of ozone depletion, water pollution, or climate change.[2]

From the perspective of the global environment, the interests of the developing and the developed worlds are inextricably inter-

Preservation and the Less-Developed World

twined. The North needs the South's contributions to reducing global environmental threats. The South needs the North's cooperation to ensure that those threats are reduced, to help relieve the maldistribution of global wealth, and to increase opportunity afforded the residents of the poorest nations. Including the LDCs in environmental preservation efforts also offers our most promising opportunity to pursue more effectively than we have in the past economic development and improvements in the quality of life of the poorest residents of the planet.

The industrialized countries of the North also have a clear and compelling interest in ensuring that the less-developed nations of the South contribute to solutions to global problems. The North also has a clear and compelling obligation to prevent or mitigate global environmental threats, since the consequences will fall more heavily on the South. The inequities in the distribution of the world's resources compel the North to take precautionary steps to avoid imposing burdens on the South that result from the benefits of economic activity that are largely enjoyed by the residents of the North. Global climate change and other environmental problems threaten all of the earth's inhabitants, but the greatest threat is to poorer countries.

The problems of environmental degradation and poverty are intricately connected. Many of the most pressing environmental problems in the developing world are the byproducts of modern, industrialized life. People in these countries depend immediately and directly on natural resources for their survival. In their struggle for survival, the poor of the world are likely to harm their environment and make their survival even more tenuous. Some have argued that we need not worry about climate change and other global threats because humans have always been able to adapt to change. However, when humans stretch resources for survival, little room is left for flexibility and adaptability. LDCs are tempted to sell off natural resources in order to raise cash to meet foreign debt. Selling fisheries to the commercial export industry, for example, not only deprives people of their livelihoods, but increases environmental damage as the pressure the dispossessed put on urban areas and agricultural lands increases.

The reliance on older, dirtier technologies, the pressure to generate foreign capital to service debt, the lack of political demands for conservation, and inadequate institutional mechanisms to protect resources are the result of choices made by the elites of less developed nations, although in most cases these choices have been constrained by international developments and conditions beyond the elites' control. These trends are not sustainable; as the population continues to grow, there will be even greater pressure on an increasingly shrinking supply of resources.

This chapter reviews some of the challenges in linking the future of the North and the South, examines some of the most prominent efforts to bring together environmental and development concerns, and assesses these efforts. It argues that the countries of the North have largely failed to recognize and commit themselves to reducing inequality and helping the countries of the South remedy the problems that plague them, even though such a commitment is essential in reducing global threats. It is in the interest of the residents of the North to commit to assisting the South in its ecologically sustainable development efforts in order to achieve global environmental protection goals.

Challenges in Implementing Global Accords in the LDCs

North-South Tensions

The tension between the developing and developed nations is considerable, with historical roots in colonialism, economic exploitation, military adventurism, nationalism, and other factors. This tension has become more pronounced as the debate over addressing global environmental problems has evolved during the past two decades. Those in the South fear that their aspirations of economic growth, reduced poverty and starvation, and improved health and education are now to be sacrificed to a global effort to reverse the environmental excesses of the wealthy nations. They worry that global efforts fashioned by wealthy nations will prevent them from harvesting their natural resources and expanding their industrial base.

Challenges in Implementing Global Accords in the LDCs

Residents of the developing world point out that developed nations have been slow to impose pollution controls on their own industries and quick to harvest their forests and other natural resources at unsustainable levels, and that global environmental problems are primarily a result of actions in the North. The United States has been singled out as the most egregious example of excessive consumption. As individuals, Americans produce more pollution and consume more energy and other natural resources than any other people on the planet. With 5 percent of the world's population, the United States produces about one-third of the world's carbon dioxide emissions from fossil fuels and consumes about one-fourth of all energy produced.[3] A Third World resident consumes about 8 percent of the energy used by a typical American.[4]

Some officials of the industrialized world have demanded that countries with tropical rain forests reduce timber harvests. Nearly half of these forests have been lost through logging, ranching, agriculture, and other development; a new acre is cleared about every second. But the export of tropical hardwoods is a major source of income for these countries. Trees are also a major energy source in the developing world. The destruction of rain forests deprives indigenous peoples of their homes and way of life, results in the extinction of plants and animals, destroys irreplaceable sources of pharmaceuticals and industrial products, and permanently scars some of the most beautiful regions of the world.[5] Japan and the United States are the largest consumers of tropical hardwoods. The United States has harvested most of its old-growth forests, which would otherwise help absorb carbon dioxide and reduce the threat of global warming. Other developed countries have done likewise, and they continue to destroy their remaining old-growth forests through acid rain, ozone, and other forms of air pollution. And they demand that the LDCs act differently.

Population and Consumption

The policy agenda is also limited by MDC interests. For example, some leaders of LDCs point out the hypocrisy of the MDCs, to whom it is acceptable to discuss coercive efforts to curb population growth in the South but not to reduce consumption in the North. The MDCs emphasize the need to reduce population growth in the LDCs, but many there believe that the carrying capacity of lands can be expanded with the introduction of a few key imports. Because large families are a protection against the threat of poverty, population reduction policies must be combined with actions to reduce that threat. Population growth in the South cannot be diminished so that consumption in the North can increase.[6]

The world's population grows by some 90 million persons a year; most of this growth occurs in the developing world. The population of some LDCs will double, if current rates continue, in the next thirty to forty years.[7] But some in the developing world argue that if all the world's citizens had the same standard of living as those living in the LDCs, global environmental problems would be much more manageable. Residents of LDCs fear that wealthy nations will demand that poorer nations control population growth, reduce the consumption of resources, and delay industrialization plans, and then fail to provide the economic assistance necessary to address the problems of poverty, disease, and suffering that afflict one-fourth of the world's 5 billion inhabitants.

Debate over environmental policy in the North has increasingly raised the issue of intergenerational equity—that we must conserve resources for use by our posterity. From the perspective of the South, however, immediate equity is a much more pressing and compelling concern; the needs of current generations are not being met, and they should be given priority over the threat of future calamities.[8] What those in the North really want, many in the South believe, is license to overconsume, instead of ensuring that everyone has the opportunity to be healthy, fed, and educated. From the perspective of the South, the problem of the North is "overdevelopment." These patterns cannot be sustained into the future without threatening all people. "Sustainable development does not just mean conserving present resources for future generations," argued one spokesperson for the Third World Network (a group of some two hundred NGOs) at the 1992 United Nations Conference on the Environment and Development; it also means

Challenges in Implementing Global Accords in the LDCs

"reducing the excessive consumption of a minority so that resources are 'freed' to meet the basic needs of the rest of the humanity within this generation."[9]

Differences in the Global Agenda

Until recently, the North has loosely defined the global commons to include the atmosphere, oceans, and Antarctica. This definition has now been expanded to include tropical rain forests and biodiversity. Efforts directed by the North to conserve these resources challenge the sovereignty of LDCs.

In traditional societies, the idea of the commons means collective ownership, equal access, and sustainable use. Some aspects of the North's idea of a global commons threaten the interests of the South. Transnational corporations' demands to control the commercialization of genetic resources challenges the traditional idea of the commons and threatens an important potential financial resource for the developing countries. The policy agenda is also limited by MDC interests. The most immediate environmental threats to the South—polluted water, air pollution in urban areas, and erosion of topsoil, for example—are not given the same priority by the nations of the North.

The Uruguay Round of the General Agreement on Tariffs and Trade is also, from the perspective of the South, oriented toward concerns of the North. Some of the talks have sought to reduce restrictions that Third World nations place on foreign companies (such as use of local materials and export limits) and to require that foreign companies in the service sector be treated no differently than domestic firms. To the developing world, the industrial world's insistence on patent protections for companies that develop biotechnology resources in the South is economic colonization. The overall implication of trade talks for the nations of the South is a limit on their ability to restrict the activities of transnational companies.[10]

The experiences of India and China in implementing the Montréal Protocol show that it is sometimes possible for countries to sidestep North-South conflict. India focused on the likely adverse consequences of a CFC ban on economic growth and the spread of consumer goods in India, and made aggressive demands for financial and technical assistance. China, in contrast, moved to support the agreement. China may have so acted for several reasons: Unlike India, it had already taken steps toward economic liberalization and growth, it was more willing to accept Western aid, its authoritarian structure was more malleable by leaders and much less open to dissenting views, its industrial sector is much more closely integrated into the political authority structure and centralized direction of the economy was well established, and it sought recognition as a global leader. Chinese institutions also saw support for the ozone protection agreement as a way to strengthen their political power domestically and internationally. Unlike their Indian counterparts, Chinese officials were able to make this a technical issue, divorced from broader North-South conflicts.[11]

Poverty in the LDCs

Many people in LDCs suffer from poverty, malnutrition, poor health, and illiteracy. Their countries are plagued with high unemployment, population growth rates that exceed the carrying capacity of the land, limited educational opportunities, and widespread disease. Many people spend their energy simply finding enough food to eat as well as fuel to cook the food and warm themselves. Many suffer from chronic problems of contaminated water and disease, and many have little access to health care. According to a recent Human Development Report of the United Nations Development Program, 10 million older children and young adults and 14 million young children die each year. Most of the forty thousand deaths that occur each day are preventable.

Some 1.2 billion people in the world go hungry every day; 1.5 billion people lack access to basic health care; 1.5 billion do not have safe drinking water; about a billion adults cannot read or write; over 300 million children of primary school age are not in school; and more than 2 billion lack safe sanitation.

These are only average figures; in some countries and communities, conditions are much worse. In the rural areas of some African countries, only 10 to 15 percent of the people have safe drinking water; in South America, the numbers range from 8 to 27 percent among the poorest nations. In many African countries, fewer than 5 percent of the people have access to sanitation services; among some Asian and South American countries, the figures are only slightly better. The life expectancy at birth for people in Africa is 51.9 years; in South America, 65.5 years; 61.1 years in Asia; but 74 years in the United States and Europe. The infant mortality rate in Africa is 106 deaths per one thousand live births. In South America the rate is 58; in Asia, 73; and in the United States, 10. The child mortality rate (deaths of children less than five years old per one thousand live births) is 163 in Africa, 78 in South America, 108 in Asia, and 12 in the United States.[12]

The distribution of income in the world is extremely unbalanced. Prospects for individuals and families whose incomes are in the lowest percentiles are even bleaker than the aggregate figures show: 77 percent of the people of the world earn only 15 percent of its income. The average income in the North is eighteen times greater than that in the South. The gap between the rich and poor countries is expanding. In 1960, for example, the difference between the per capita income of the industrial and the developing countries was $5,700; in 1993, the difference had almost tripled to $15,400.[13] Unemployment rates in some parts of the world exceed 50 percent, dwarfing the levels of 15 to 20 percent reached during the Great Depression in the United States and Europe. The per capita income in some African nations is about $150 per year; in some South American countries, it is less than $500 per year; and in the poorest Asian nations, less than $250 per year. In contrast, the per capita income in the United States is more than $18,000.[14]

The 1996 Human Development Report published by the United Nations Development Program emphasized the increasing gap between the rich and poor in the developing countries. In Brazil and Guatemala, for example, the average income of the poorest fifth of the population is only about ten percent of the average per capita income in the nation. In other nations, such as India and Indonesia, the division of national wealth has been more egalitarian, with an average income for the poorest fifth of 44 percent of the average per capita income. Many Asian countries have invested in health and education, microcredit programs aimed at poor families and entrepeneurs, and improving the status of women. These countries have also pursued national economic growth through market-oriented economic policies, and the results have been among the most dramatic improvements in economic condition ever recorded in history.[15]

Women and Global Poverty

Women and children bear much of the brunt of poverty in the world. Women suffer from maternal-related deaths, respiratory disease (from inhaling the smoke of cooking fires), and other maladies that result from poverty. More than half a million women die each year from causes related to pregnancy and childbirth.

Wide social disparities exist between men and women. Much of the work done by women is underpaid and undervalued. Women's wages remain at two-thirds those of men. Women are often prevented from owning property.

Primary school enrollment of girls is only a little over half that of boys. Female literacy is only about two-thirds that of males. The adult female literacy rate is as low as 10 percent in some areas of Africa, 65 percent in South America, and less than 10 percent in the poorer nations of Asia. The rate for males is generally higher, but it still falls well below U.S. and European levels.[16]

Discrimination, which is manifest in land ownership laws, employment practices, education, and development programs, contributes to the global feminization of poverty. Because of the low status of women in most societies, development projects often ignore their role in local economies. In fact, women make up 50 percent of the agricultural labor force and more than 70 percent of some industries such as clothing manufacturing, but their efforts are largely outside the formal economy. Women have few legally enforceable rights concerning land ownership and control of financial resources. In many areas, only a small part of a man's income goes to the family, so development projects aimed at enhancing male earn-

Challenges in Implementing Global Accords in the LDCs

ings may do little to alleviate family poverty. In rural areas, women often work from ten to sixteen hours a day; their primary responsibility is to produce the food on which their families depend. Men are often prohibited, by "custom," from sharing in the workload.[17] Government policies that mandate low prices for basic foods and promote the production of cash-producing export crops make it even more difficult for women to grow enough food to feed their families. Women are effectively prohibited from participating in irrigation projects because they cannot own land or lack access to credit. The distribution of mechanized farm equipment heavily favors men and the production of cash crops, rather than helping women produce food for their families. Women often lack access to new seeds, fertilizers, and other technologies that could lighten their workload and increase their output. Development projects are evaluated in terms of narrow indicators such as export revenue or employment in the formal economy, which largely exclude the contributions and concerns of women. As a result of these and other policies, Africa's per capita grain production today is lower than in 1960.[18]

The role that women play in national economies is usually greatly underestimated. The number of women working in the informal sector is not reflected in official labor market data; in some countries, more than 50 percent of the urban labor force is employed in the "shadow economy." Women may constitute more than two-thirds of the shadow economies of the less-developed nations.[19]

Throughout the world, millions of women and children are employed in the shadow economy. Hired to do piecework in their own homes, they weave carpets, stitch apparel, roll cigarettes, and launder clothes. They do not enjoy stable work and pay; they are not given health benefits; and they are often exploited by employers. Theirs is a bleak, harsh struggle for existence: "Lacking the tools to achieve self-sufficiency—literacy and arithmetic, loan credits, and marketing skills—these women tread a circular path of desperate job seeking and miserable pay."[20]

Piecework is usually the employment opportunity of last resort for women who are poor, lack skills, and have child care responsibilities. Although piecework may provide desperately needed income, it rarely translates into an enhanced status for women: the pay is so low that it largely reinforces women's second-class status.[21] One anthropologist described piecework laundering as particularly exploitative: It involves women in an exhausting cycle of "locking up their children, traveling across town to collect clothes from a middle-class household, bringing it [sic] home to wash, and then repeating the journey to return the laundered goods."[22]

When women are employed in the formal economy of the third world, they do not fare much better. Much as are women in the more-developed world, they are concentrated in low-paying manufacturing and service jobs. In addition to working twelve-hour days, at home they still face housework, child care, and other family responsibilities. Women lack the resources or the time (or both) to obtain adequate health care, and their work is often hazardous, exposing them to toxic chemicals, particulate pollution, and other threats. Employers can threaten to turn work over to pieceworkers in response to employees' demands for better conditions and compensation.

Lack of Effective Governmental/Regulatory Infrastructure

Many LDCs lack the governmental, regulatory infrastructure to develop effective regulatory programs. They lack effective governmental institutions and a tradition and culture of compliance with regulatory requirements. They lack the necessary scientific infrastructure to understand the problems they confront.

Development programs devised by the North have often emphasized large-scale, politically visible projects that impose western technologies and ways of thinking on cultures and societies that are organized and structured much differently. Relief or short-term assistance is often not integrated with long-term, sustainable interventions that promote self-sufficiency. A recent UN study of the developing nations argued that these nations must "get their house in order." They must address such issues as "rapid population growth, corruption, inadequate technical and managerial standards of

Challenges in Implementing Global Accords in the LDCs

competence, extreme income disparities and 'unrealistic' official rates of exchange."[23]

Many Third World countries spend a large percentage of their resources—some as much as one-third of their budgets—on weapons. Despite the tremendous reduction in East-West tensions, arms imports by Third World nations continue to grow, diverting resources from human and environmental concerns and producing, in turn, the ecological damage that accompanies military activity.[24]

An extreme manifestation of this problem is the lack of stable government, a result of internal strife. The United Nations Food and Agricultural Organization recently warned that some 35 million people are at risk in the six African countries most threatened by famine: the Sudan, Ethiopia, Somalia, Liberia, Mozambique, and Angola. All have been plagued by civil war. War and related tensions prevent food supplies from getting to those in need. Wars and other forms of military activity pose tremendous environmental challenges. According to human rights groups, the International Red Cross, and the UN, there are from 65 to 100 million undiscovered live land mines throughout the world. If all the income of all the people of Cambodia were dedicated to disarming landmines within its borders, it would take two years. Land mines continue to be laid faster than they are removed.[25]

Relief agency officials have observed that "there has never been a time in recent history when so many disasters with such different natures have struck so many people."[26] Many countries, others note, "continue to mismanage their economies—tolerating inefficiency for political ends, squandering precious resources on wars, [and investing] too little in education."[27] Many countries continue to pursue protectionist policies, licenses and regulations on entry into industries, and price controls that stifle economic activity.

Lack of Investment Resources

The LDCs require massive amounts of capital to upgrade their industrial infrastructure. Since newer technologies are generally less polluting, economic prosperity can also remedy environmental degradation. Increased industrial activity, however, also increases emissions.

Until 1983, the net flow of funds (assistance and loans) from the North to the South averaged about $40 billion a year; the net flow reversed direction in 1983, and the poorer countries throughout that decade sent more than $20 billion more per year to the wealthy nations than they received in new assistance. In 1994 the LDCs owed some $1.8 trillion dollars to banks and international lending institutions. The debt-GNP ratio—the dollar value of outstanding medium- and long-term debt as a percentage of GNP—is nearly 40 percent in the developing world, and as high as 80 percent in the poorest African nations.[28] In sub-Saharan Africa, debt service payments are $10 billion a year, four times the annual amount these countries spend on health and education.

Restructuring the debt of many countries in the 1980s eased somewhat their financial burdens. As commercial banks have reduced the debt they hold from the poorest countries in favor of debt to East Asian countries and private sector lending, the poorest nations have increasingly turned to such multilateral lending institutions as the World Bank. The share of debt of the lowest-income countries to multilateral institutions grew from 15 percent of total debt in 1980 to 24 percent in 1992. The multilaterals have restructured the debt of middle-income countries, but did not provide relief to the poorest countries until 1996.

In 1995, the Group of Seven major industrial nations asked the World Bank and the International Monetary Fund to devise a program of debt relief for poor nations. This was a response to studies such as one issued by Oxfam International, which found countries such as Uganda spending $3 a person a year on health care and $17 a year on servicing international debt. In 1996, the Bank and the IMF proposed a plan to reduce the debt of twenty of the world's poorest and most indebted nations, including Zaire, Zambia, Uganda, Madagascar, Tanzania, Nicaragua, and Bolivia. These countries would first undergo a three-year economic reform program outlined by the Bank and the IMF and would then be eligible for forgiveness of up to two-thirds of the debt they owe to

Challenges in Implementing Global Accords in the LDCs

the Paris Club. (The Paris Club is a body official creditors that represent the major industrialized nations in decisions affecting the debt owed by developing countries; Club members hold 42 percent owed by developing countries; the World Bank, IMF, and other multilateral institutions hold 21 percent; other bilateral creditors such as Arab countries and the former Soviet Union nations also hold 21 percent of the debt, and 16 percent of the debt is owed to private creditors.[29]) After another three years, up to 90 percent of their debt could be forgiven; the Bank would pay off some of the debts owed to multilateral lending institutions, and the IMF would lend money at a low interest rate (one-half of one percent) to the countries to pay off some of their debt. Their debt would eventually be no greater than 350 percent of their annual export earnings. The proposal was approved by the Group of Seven in September 1996, the first time debts to international lending agencies had been forgiven and the largest reduction in other indebtedness ever agreed to. But critics argued that the six-year plan will take too long to reduce the crushing burden of debt and does not reach enough needy countries.[30]

Although the debt crisis may be over for many countries, it is not for the poorest ones.[31] Tables 6.1 and 6.2 describe in more detail the daunting debt challenge that confronts the developing world.

Environmental Damage and Poverty

Global climate change and other environmental problems threaten all of the earth's inhabitants, but the greatest threat is to poorer countries. Alan Durning has written, "poverty can drive ecological deterioration when desperate people overexploit their resource base, sacrificing the future to salvage the present." Environmental decline "perpetuates poverty, as degraded ecosystems offer diminishing yields to their poor inhabitants. A self-feeding downward spiral of economic deprivation and ecological degradation takes hold."[32]

Agricultural and environmental problems are also closely interconnected. Billions of tons of fertile soil are lost each year because of the cultivation of steep lands, reduced forests and vegetation, and

TABLE 6.1

THE GROWTH OF EXTERNAL DEBT, 1982–1990
(U.S. $ BILLIONS)

Category	1982	1983	1984	1985	1986	1987	1988[a]	1989[b]	1990[c]
Total debt, DRS countries[d]	753	819	856	952	1047	1176	1156	1165	1189
Long-term debt	561	644	684	780	882	999	980	995	1039
Official sources	199	221	234	296	360	440	443	467	517
Private sources	362	423	450	484	522	559	537	528	522
Short-term debt	168	141	134	132	122	135	141	139	127
Use of IMF credit	24	34	36	40	43	43	35	32	23
Total debt, other developing countries	86	86	81	89	99	116	128	125	130
Total external debt, all developing countries	838	905	936	1041	1146	1292	1284	1290	1319

[a] Estimates
[b] Preliminary estimates
[c] Projections
[d] Countries reporting to the World Bank Debtor Reporting System

SOURCE: Morris Miller, *Debt and the Environment: Converging Crises* (New York: United Nations Publication, 1991), 11.

improper irrigation. Soil erosion also shortens the life of dams and irrigation projects, fills in canals and harbors, and harms wetlands.

Wood is the primary energy source for half of the world's inhabitants; forests have been cleared for farming, and animal dung and crop residues are used for fuel instead of to build soil.

Increased agricultural output has relied on energy and mechanization as well as irrigation, chemical fertilizers, pesticides, and herbicides, all of which pose significant environmental threats. Excessive irrigation wastes water, depletes aquifers, and washes nutrients from soil. Waterlogging and chemical reactions damage soil. Fertilizers and pesticides contaminate drinking water and harm waterfowl. Habitat destruction threatens the loss of genetic diversity.[33]

TABLE 6.2
LONG-TERM AND FINANCIAL FLOWS IN DEVELOPING COUNTRIES, 1982–1991
($ U.S. BILLIONS)

Long-term debt and financial flows	1982	1983	1984	1985	1986	1987	1988	1989	1990	1991
Debt disbursed	562.5	644.9	686.7	793.7	893.9	996.3	1121	1143	1190	1230
Disbursements	116.9	97.2	91.6	89.3	87.7	86.7	103	93	110	114
(from principal credit)	84.6	65.0	58.9	57.8	50.8	48.5	—	—	—	—
Debt Service	98.7	92.6	101.8	112.2	116.5	124.9	152	145	146	152
Principal Repayment	49.7	45.4	48.6	56.4	61.5	70.9	79	75	75	77
Interest	48.9	47.3	53.2	55.8	54.9	54.0	73	70	71	75
Net flows	67.2	51.8	43.0	32.9	26.2	15.8	25	19	35	37
Net transfers	18.2	4.6	−10.2	−22.9	−28.7	−38.1	−45	−51	33	−42

SOURCE: Morris Miller, *Debt and the Environment: Converging Crises* (New York: United Nations Publication, 1991), 14.

PRESERVATION AND THE LESS-DEVELOPED WORLD

The debate over adequacy of global food production also focuses attention on the challenges facing the developing world. Food production increased by 300 percent between 1965 and 1990. Optimists argue that yields are still increasing and will continue to do so with new investments in agricultural research and technologies and a continued shift toward freer agricultural markets. One scientist summarized the cornucopian view: "there's no way to predict which new advances will have the most impact, but we've never invested money in agricultural research and not gotten a strong return." Scientists anticipate new breakthroughs in fertilizers, seeds, and other ways to boost production. Pessimists argue that it will be extremely difficult to duplicate past increases as pollution from fertilizers and pesticides build up, fertile land becomes more scarce, and the price of food increases. "Food scarcity may become the defining issue of our era," warns Worldwatch head Lester Brown. Optimists reply that increases in food prices pose no global economic problem, since food commodities make up only 2 percent of all global costs. But for people who live on the equivalent of $1 a day, even small increases in the price of food may increase starvation.[34] If agricultural output is to increase, new technologies will need to be transferred to farmers along with assistance to increase their access to water.

In the developing world, environmental problems threaten human survival: Food shortages, disease, and death affect millions each year. Developing countries rely heavily on inefficient technologies that have been discarded as obsolete and exported by industrial countries.[35] The factor that underlies everything else is population growth. As Garrett Hardin wrote twenty-five years ago,

> The pollution problem is a consequence of population . . . . as population became denser, the natural chemical and biological recycling processes became overloaded . . . . Freedom to breed will bring ruin to all.[36]

Even if food production trends can be reversed, and enough food is produced, civil wars, ineffective government, inefficient distribution means, and other social constraints will not likely be overcome. Because of the earth's spiraling population and increasing economic output, the scarcity of renewable natural resources will likely increase; this may, in turn, trigger civil and international strife.[37]

Options for Assisting the LDCs in Implementing Global Agreements

Studies, reports, statements by policy makers, and international environmental agreements themselves all emphasize that a major commitment of financial assistance from the North to the South is essential in implementing these accords. Most of the attention has focused on two sets of options: technology transfer programs to upgrade industrial and energy facilities and establish regulatory programs and scientific support agencies, and provision of new resources from the North to the South to fund implementation of global accords and ecologically sustainable development. The assistance promised and delivered thus far by the North to the South can be assessed from two perspectives. First, is the current framework of assistance sufficient to provide for implementation of global agreements? Second, even if the assistance is sufficient to achieve the main objectives of existing agreements, is it nevertheless inadequate to meet the needs of the less-developed world?

Technology Transfer and Institutional Capacity

New, cleaner technologies play a key role in integrating the environment and development agendas. For instance, the two primary means of reducing emissions of greenhouse gases are to replace traditional fossil fuels with cleaner ones, and to improve the efficiency of energy use. Since the technologies for alternative energy sources and conservation have been and will likely continue to be produced largely in the developed world, those nations will need to facilitate the transfer of these technologies to the developing nations.

Many leaders of the developing world fear that these new technologies will be imposed without sufficient sensitivity to local conditions and concerns. The cultural context in which new technologies are employed are at least as important as, and probably more important than, the technologies themselves.

Leaders of developing countries also fear that companies in the North will reap great profits because they own the intellectual property rights to such technologies. Since patents usually expire in less than twenty years, and since many conservation technologies are in the public domain, the problem of intellectual property rights has been a relatively minor one in practice.[38]

Developing countries are often unaware of new technologies and lack the technical infrastructure to take advantage of them. Some technological investments have resulted in tragedy, such as the release of deadly chemicals in Bhopal, India. Others have been poorly suited to local environmental and cultural conditions.

Energy efficiency is one of the most important investments developing countries can make. According to one estimate, these countries will need to invest $70 billion a year over the next several years, and twice that much in the first years of the twenty-first century, in order to meet growing energy demands. But that amount is about one-fourth of all the money they will have available for capital investments, and money spent on energy will not be available for other pressing needs. Using energy more efficiently is cheaper than increasing energy capacity. However, developing countries have tended to do just the opposite. China, for example, purchased old refrigerator factories from the industrial countries in the late 1980s in order to increase the number of refrigerators available to households in Beijing. The inefficient appliances increased electricity consumption, requiring the construction of new power plants and increasing emissions of greenhouse gases from coal-fired power plants; increased power outages; and released CFCs. Asian countries cannot keep up with the demand for air conditioning and refrigeration among their burgeoning populations. They have not been given much help from international lending organizations: Energy efficient projects represent only about 1 percent of lending by the World Bank.[39]

No single model of energy efficiency can be applied to the entire developing world; technology must be compatible with local practices and opportunities. Six factors appear to be critical in successfully transferring technologies to the developing world:

1. Local demand for the technology and a willingness to make use of the new equipment or hardware and learn how to operate it;

2. Local accessibility to information concerning new technologies;

3. Supporting infrastructure, including education and training, transportation, and communication;

4. Clear contribution of the technology to the strengthening of the local community, with minimal governmental subsidy required;

5. Availability of initial capital; and

6. Appropriateness of the technology in meeting local needs and conditions.[40]

In 1976, the U.S. EPA began to review the environmental consequences of foreign aid projects. The Agency for International Development has funded a number of actions by EPA officials to improve environmental conditions in recipient nations. EPA assistance to less developed countries seeks to contribute to overall efforts to address global environmental problems, contribute to the stability of foreign governments, and help countries develop their own institutional capacity, including effective legal systems. One example of this latter kind of effort is EPA assistance to Russia that includes demonstration projects for formulating environmental regulations. Russian environmental laws include some very strict, aggressive provisions, but they are largely ignored and have little impact on industrial practices. The demonstration projects create teams of Russian government lawyers and other environment specialists, industry representatives, local government officials and others who work with U.S. consultants to negotiate practical, enforceable, regulations. Transferring these cultural and legal concepts from one country to another poses tremendous challenges that dwarf the difficulties in technology transfer but are nevertheless essential steps in making regulation more effective.[41]

The exodus of scientists and engineers from LDCs has been recognized as a problem for decades. In the mid-1970s, UNDP began a technology-transfer program entitled Transfer of Knowledge Through Expatriate Nations (TOKTEN). Some forty countries have participated. Nonresident professionals work as volunteer consultants to host-country organizations in order to strengthen research activities, explore alternatives for products and processes, and otherwise help address environmental problems. In India, for example, the program began operating in the early 1980s, and, by 1994, had brought more than 550 visits to India by expert consultants to some 250 organizations. The initial effort focused on providing assistance to Indian efforts to build a research and development infrastructure; later visits sought to link R&D with the development of new technologies and other assistance to industries so they can become more competitive in global markets and more environment friendly.[42]

The UNCED report recommended a number of ways to expand the authority of the UN institutions in the field of environment and development:

> existing international institutions at the global and regional levels in the field of environment and development, including UNEP and UNDP, should be adapted to changed circumstances in order to support sustained development;
>
> among the goals of institutional reform at the global and regional levels should be an enhancement of the capacity of institutions at the national level, especially in developing countries, to ensure the full integration of environment and development;
>
> states should promote and support, in more concrete ways, the effective participation of developing countries in the negotiation and implementation of international agreements or instruments—such support should include assured financial assistance to cover the necessary travel expenses to meetings and access to the necessary information and scientific-technical expertise on preferential terms; and

Options for Assisting the LDCs

assist developing countries and economies in transition in their national efforts at modernizing and strengthening the legal and regulatory framework of governance for sustainable development, having regard to local social values and infrastructures.[43]

Bilateral and multilateral donors can help developing countries build up their social and physical infrastructure. Regional assistance programs in particular are promising.

Global institutions can enhance the capacity of national institutions, especially those of developing countries, to implement global agreements. Technical assistance is critical in developing the scientific and regulatory infrastructure necessary to address environmental problems. International organizations can make an important contribution by working out these differences so that LDCs can share in the benefits of harvesting their natural resources. Information concerning alternative policy tools, such as decentralized, marketlike regulatory approaches could also be fruitfully disseminated.[44]

Finally, we also need to change the way we assess national economies. Current economic analysis, Robert Repetto argues, ignores the loss of environmental and natural resources in its assessing of economic progress in Third World countries; economists today concentrate on only two kinds of assets: human resources and invested capital, treating natural resources as some sort of "freebie." If a country's balance sheet shows the income gains made from the timber industry, then it should also account for the loss in the forest's natural resources to show the *real net* gain from production. However, although Third World nations are usually the most dependent on their natural resources, they "use a national accounting system that almost completely ignores their principal assets . . . A country can cut down its forests, erode its soils, pollute its aquifers and hunt its wildlife and fisheries to extinction," Repetto warns, "but its measured income is not affected as these assets disappear. Impoverishment is taken for progress."[45] This approach eventually results in substantial depletion of natural resources, which will ultimately bankrupt these developing countries.[46] They are forced to take whatever prices are offered by the large transnational corporations that dominate international trade and are not in a position to raise prices to encourage conservation or account for the true costs of exploiting limited resources.[47]

Funding Implementation of Environmental Agreements and Ecologically Sustainable Development

DEBT SWAPPING

Much more difficult than the transfer of technologies and institutional capacity building is the transfer of resources to purchase newer technologies. The LDCs require massive amounts of capital to upgrade their industrial infrastructure, but their indebtedness is a major barrier.

In 1990, U.S. Treasury Secretary Nicholas Brady proposed that banks provide some debt relief for the middle-income debtor nations. In 1992, attention shifted to reducing debt owed to U.S. taxpayers. The Bush administration proposed that billions of dollars of Egypt's debt to the United States government and 70 percent of the money Poland owes to the United States be forgiven. But some fear that providing debt relief on government claims might discourage the "self-help efforts of indebted countries working diligently to repay their debt." If private creditors pull out and governments provide debt relief instead of more credit, the funds available to the LDCs might decline.[48]

In debt-for-equity swaps, corporations purchase part of a nation's debt from banks and then trade it for an asset such as a state-owned factory or utility. In debt-for-nature or debt-for-development swaps, an environmental or nongovernmental group typically buys discounted debt from a bank and then restructures it with the debtor nation: payments are made in local currency and then invested in conservation projects such as a national park or public health program. Although debt swaps represent only about 2 percent of debt conversions, in some countries they have made a major difference and could make more if expanded.[49] Madagascar, for example, has cut its debt to commercial banks in half through debt-for-nature trades.[50] Debt-for-nature swaps have been successful in Bolivia,

Options for Assisting the LDCs

Ecuador, Costa Rica, the Philippines, Madagascar, and Zambia. In Bolivia, for example, Conservation International bought $650,000 of Bolivian debt for $100,000. The Weeden Foundation provided the cash, and Conservation International then canceled the debt; in return, Bolivia agreed to protect and expand the Beni Biosphere Reserve, home of thirteen endangered species of animals, five hundred species of birds, and the Chimane Indians. But the remaining Bolivian debt is still $4.1 billion. The size of the Third World debt and the resultant interest payments means that such swaps, unless dramatically expanded, will only have a miniscule effect on indebtedness in the developing world.[51]

GRANT FUNDING

The World Bank has promised to increase its environmental capacity through environmental assessments of bank projects and spending on environmental improvements. One important breakthrough occurred when the wealthy nations agreed, as part of the 1990 London meeting concerning the Montréal Protocol on Substances that Deplete the Ozone Layer, to contribute to a fund designed to assist the poorer nations' transition to substitutes for CFCs. A $240 million fund was created to help the LDCs that signed the protocol implement its provisions.[52] As part of the preparation for the 1992 United Nations Conference on the Environment and Development, the Global Environmental Facility (GEF) was created in 1990 to help LDCs address the threat of global climate change and other environmental problems (see Chapter 4).[53]

In June 1996, the World Bank's environment portfolio actually reached $1 billion through grants for ninety-nine projects in fifty-nine countries. Following are the percentages of projects funded in 1996 that are aimed at the major environmental problems:[54]

Mitigate climate change	33%
Conserve biodiversity	29%
Phase out ozone-depleting substances	26%
Protect international waters	11%
Projects that cross sections	1%

PRESERVATION AND THE LESS-DEVELOPED WORLD

The Global Environmental Fund can serve as a model for other efforts to provide financial incentives to encourage LDCs to participate in global agreements and assist developing countries in their efforts to modernize their economies and address poverty. But this fund's limited resources are not enough to meet current needs.

LEVELS OF AID

A World Resources Institute compilation of estimates for funding global environmental needs identified more than $110 billion in annual expenditures to address the following problems:

Population: $4.5 billion per year is needed, according to the UN Population Fund, to help the less-developed world.

Desertification: $2.4 billion per year is required to implement a plan of action to combat desertification by the year 2000.

Tropical deforestation: $8 billion per year for five years would stabilize tropical forest loss.

Fresh water: $28.2 billion per year is needed to ensure adequate water and sanitation services to 90 percent of the people of the world by the year 2000, according to a UNICEF estimate.

Protecting biodiversity: $1 billion per year would establish new biodiversity protection efforts.

Ozone depletion: $16 to $24 million per year would assist the LDCs in complying with the Montreal Protocol to protect the stratospheric ozone layer.

Climate change: Estimates to arrest the threat of global climate change vary tremendously, depending on the level of greenhouse gas emission reduction that is to be achieved. A reduction of 5 percent in the year 2005 from projected levels might cost as much as $67 billion a year ($18 billion for the OCED nations; $22 billion for the LDCs, and $27 billion for the Eastern European states).[55]

Options for Assisting the LDCs

These needs cannot easily be separated from development needs. Estimates from the World Bank and the WorldWatch Institute for development efforts throughout the 1990s project an annual budget of $125 billion per year. The overall costs of concessional international financing required to achieve the goals outlined in Agenda 21 could require another $125 billion a year.[56]

These estimates dwarf current spending on development. UN estimates for development assistance in 1990 identified the following expenditures:[57]

Source	$ Billion, 1990
Bilateral disbursements	49.1
Multilateral disbursements	12.3
U.N. sources: $8.5 billion	
Concessional loans: $3.8 billion	
Total official development assistance	61.4
Other major sources:	
World Wood Program	1.0
Special Drawing Rights from the IMF	13.9
World Bank loans (FY 1991)	16.4
International Development Association credits	6.3
Regional development bank loans	9.17

The International Development Association (IDA), the window of the World Bank where "soft" or interest-free loans are made to the poorest nations, was replenished in 1992 with $18 billion; a total of $22 billion was available from July, 1993 to July, 1996.[58] The Bank has formulated a list of priorities for environment and development projects that would cost $75 billion a year. In contrast, the MDCs currently provide about $57 billion annually in official development assistance. Of this $57 billion, 28 percent takes the form of multilateral aid; the balance is direct bilateral transfers. In addition to official assistance, $55 billion from private sources (including $28 billion in foreign direct investment and $5 billion from nongovernmental organizations) and other contributions go to the less-developed world, but much of this money does is not spent on development-related needs.[59]

The level of aid as a percentage of GNP has been relatively constant for more than two decades. With the end of the cold war many hoped a reduction in military spending could generate $100 to 300 billion a year in the North and $30 to $40 billion in the South. But continuing tensions and conflicts throughout the world have made that unlikely. Official development assistance is funded on average at 0.33 percent of the GNP of donor states. Raising that figure to 0.7 percent, a long-standing goal, would finance most of the needs cited above. Many nations reaffirmed their commitment to that goal at the Earth Summit, but others, such as the United States, refused to make any firm promises. The level of aid increased by an annual average rate of 2.4 percent throughout the 1980s as many countries significantly increased their contributions. Only three nations, the United Kingdom, Belgium, and New Zealand, decreased aid.

The western and northern European nations and Canada have focused on sub-Saharan Africa, whereas the United States gives most of its aid to the Middle East, and then to sub-Saharan Africa and Central America.[60] Table 6.3 below shows the amount of assistance provided in 1991 as a percentage of gross national product.

Another option is to redirect the $90 billion in foreign direct investment by private sources that occur each year. World Bank Studies have encouraged investments that go beyond traditional areas of economic and industrial output to build human capital through education, and social capital through strengthening local institutions and the capacity for collective problem solving.[61] The World Bank's 1996 conference on environmentally sustainable development focused on how to influence private lending in ways that contribute to Agenda 21's goals, but few specific recommendations surfaced beyond simply appealing to investors to look at the long-term implications of their actions and encouraging them to invest in education, infrastructure, basic human services, and the institutions of civil society.[62]

Options for Assisting the LDCs

TABLE 6.3
NET OFFICIAL DEVELOPMENT ASSISTANCE, 1991

Country	U.S. $ billions	Country	U.S. $ billions
United States	11.26	Norway	1.18
Japan	10.95	Spain	1.18
France	7.48	Australia	1.05
Germany	6.89	Finland	0.93
Italy	3.35	Switzerland	0.86
United Kingdom	3.25	Belgium	0.83
Canada	2.60	Austria	0.55
Netherlands	2.52	Portugal	0.21
Sweden	2.12	New Zealand	0.10
Denmark	1.20	Ireland	0.07

Country	Percent of GNP	Country	Percent of GNP
Norway	1.14	Switzerland	0.36
Denmark	0.96	Austria	0.34
Sweden	0.92	Japan	0.32
Netherlands	0.88	United Kingdom	0.32
Finland	0.76	Portugal	0.31
France	0.62	Italy	0.3
Canada	0.45	New Zealand	0.25
Belgium	0.42	Spain	0.23
Germany	0.41	United States	0.2
Australia	0.38	Ireland	0.19

SOURCE: World Resources Institute, *World Resources 1994–95* (Washington, D.C.: WRI, 1994), 227.

Environmental and other taxes could raise some funds for sustainable development, but there is little support for them now. Increases in aid would also have to be coupled with decreases in LDC debt, since some nations pay more to service their debt than they receive in new assistance, and those who are meeting payments often must sell off natural resources more quickly than they can sustain them. Reducing the trade barriers imposed by the industrialized countries could also facilitate development in the South. But opening trade also poses some challenges to the developing nations in protecting their natural resource base for future development.[63]

Prospects for Ecologically Sustainable Development

The 1991 UN Development Program argued optimistically that development works, and that judged by the basic indicators of human development, development efforts have been reasonably successful:

> Developing countries have achieved in 30 years what it took industrial countries nearly a century to accomplish. The income disparity between the North and the South is still very large: the South's per capita income is only 6% of the North's. But in human terms, the gap is closing fast. . . . True, the past record of the developing world is uneven—between various regions and countries, and even within countries. True also, the closing of human gaps between the North and the South has been witnessed only in the basic human indicators (such as life expectancy, adult literacy and infant and child mortality) and not in higher levels of education and health care, or in science and technology.[64]

Development efforts can make a difference in the lives of people in the South. If priorities are established and development is linked to things the wealthy countries want, then life can significantly improve for the billion people for whom it is now a daily struggle merely to survive. These short-run efforts to improve life in the developing world can also reduce, in the long run, the threat of global climate change. International development cooperation *has* made a difference. And the remaining agenda of human development *is* manageable in the 1990s if development priorities are properly chosen.

In the dominant Western model of development, technological innovation has played a central role. The Western model has regarded cultural and social differences as largely irrelevant in developing universal technologies to modernize rural life and economic and agricultural production. Development is fundamentally about state building—industrialization, modern education, and transportation and communications systems that would "produce an ever-increasing standard of living for the population as a whole, and a democratic, stable nation-building process would be initiated."[65] Prospects for modernization are largely in the hands of elites in developing countries because the majority of residents remain "tied to the rigid, diffuse, and ascriptive patterns of tradition."[66] The Western, modernized world is the ideal to which the developing world is to aspire; the rational, democratic, secular, rich North is to be emulated by the poor, dependent, illiterate, emotional, superstitious South.[67] Principles of political economy, well developed in the North, could be imported in the South to assure that scarce natural resources were optimally developed and that economic efficiency guided national policy.[68]

This dominant model has been widely criticized. Dependency theorists argued that the North has manipulated development to keep the South from modernizing, because that would threaten the supply of cheap natural resources and other benefits. Modernization was still the goal, but to accomplish this, the North was to leave the South alone to develop industrially, democratically, and technologically.[69] Dependency theory offered a strong critique of traditional development, but its prescriptions, apart from ensuring independence for the South, failed to provide an alternative to Western technology, industrialism, bureaucracy, and social structures.

Other critics argue that development efforts have largely been ineffective in bringing about substantial change or materially improving conditions among the poorest people of the third world. Much of the wealth that is generated remains in the hands of elites in the recipient countries and among the holders of project contracts in the donor nations. Because intervention disrupts cultures and traditions that have been central to the lives of native peoples, it has profound consequences for their existence. Development efforts have reinforced authoritarian governments and discouraged the recognition of the political and land rights of indigenous peoples. Development projects often involve large-scale harvesting of natural resources, thus disrupting the lives of peoples who rely on those resources and encouraging rural populations to migrate to urban areas, where they are likely to become part of a dependent urban underclass.

The current approach to development has many other problems. Development officials have become almost a parody of ineffectiveness. Development experts are often unwilling to travel to sites that are far from major transportation corridors; they prefer more attractive, accessible sites. Projects are often carried out in areas that have the best soil and most plentiful water, while areas that lack these conditions are often ignored. Development experts often limit their interaction and instruction to men and fail to recognize the primary role women play in food production and family health, nutrition, and literacy. Projects usually fail to reach those who have the least power to obtain available help. Many projects provide short-term relief rather than long-term development; other efforts provide instruction but fail to help develop the social organizations and networks necessary for real self-sufficiency. Outsiders impose projects, rather than native peoples selecting them. Bureaucratic mandates leave little room for flexibility. Politically visible, large-scale projects are the center of development; these are supported by corporate lobbyists who are also contractors. Profits from enterprises do not usually stay in the recipient country but accrue to these corporations, often increase the international debt of recipients. Corruption and malfeasance by recipient governments are perceived to be widespread. Agricultural programs emphasize export crops instead of improving the domestic food supply.[70]

Many development efforts are ultimately aimed at producing saleable products that will generate revenue, rather than helping people become more self-sufficient or improve their quality of life. Development projects need to focus on basic survival needs, but assistance must also be given to community organizations and resources that will lead to increased self-reliance and the capacity to respond to new problems.[71]

Options for Assisting the LDCs

Perhaps most significant has been the growing awareness of the social and ecological consequences of development projects and the unsustainability of exploiting natural resources. Social disruption has destroyed the knowledge and practices that permitted people to live in harmony with their environment for hundreds or thousands of years. New technologies may ultimately damage natural resources, reduce the food supply, and disrupt cultural practices that are central to community life.

Ecologically Sustainable, Grassroots Development

Dissatisfaction with the traditional approach to development has stimulated discussion about the prospects for alternative approaches. The basic elements of an alternative model of development have been sketched out by both theorists and practitioners. These elements differ significantly from those of the traditional model. They are united in that they embrace diversity and call for different paths to development. Development is primarily a function of the interests and desires of local residents, who identify their problems, propose solutions, and work alongside those who can provide external technological and financial support. Crops are produced primarily for domestic consumption rather than export. Sustainability and development are emphasized along with environmental concerns. These efforts are sensitive to how change affects traditional cultures and to the need to reinvent community and individual identity in response to change. They provide opportunities for native peoples to examine their cultures and to reconcile traditional beliefs and practices with the imperatives of change. They respect the energy of culture, which can "energize a people to renew themselves and their society," "mobilize individuals, groups, and communities to a heightened sense of purpose," and help "people to reach within themselves to find a previously hidden reservoir of strength and resolve." Cultural energy grows with use in awakening the "latent power" of individuals and groups.[72]

To be comprehensive, development efforts must incorporate a wide range of solutions. They must address health care, education, literacy, food production, environmental quality, and other considerations. Development needs to include efforts to reduce population growth lest it overwhelm other efforts to produce food and preserve ecological systems. Perhaps most important, development should take place *through* people, not to them or for them, as it reinforces local cultures and traditions.

Technology plays a central role in this model of development. The emergent model of integrating culture and development promises to ensure that the technologies employed are compatible with and serve to strengthen local cultures and practices. Most modern technologies used in development projects are dependent on fossil fuels and other limited resources, are poorly integrated with natural processes, are large in scale, and are justified by relying on a relatively narrow concept of efficiency and rationality. Alternative or soft technologies generally are adapted to natural cycles, minimize pollution, use renewable matter and energy, are labor intensive, minimize per capita use, take long-term costs into account, are smaller and simpler and less dependent on a specialized technical elite than hard technologies, and adapt to societal needs and ecological constraints. They often rely on materials that combine old and new technologies. Perhaps most important, they reject the idea that society must adapt to technological developments; instead, societal concerns shape technologies and collective decision making controls them.[73]

One ambitious effort to integrate culture and development was aimed at the Kuna Indians of Panama. In the early 1980s, the Kuna created a park with the help of outside funding sources. The park was designed to facilitate the study of Kuna traditional knowledge of sustainable agricultural practices in tropical rain forests and to merge this knowledge with Western scientific knowledge. Kuna elders were to work alongside university-trained scientists in identifying plants and studying traditional ways of protecting the forests.

Intergenerational differences among the Kuna over the nature of their culture have complicated efforts. It is not clear whether the old culture will be renewed and protected. Part of the problem is a logistical one, since both the elders and the scientists live several

Options for Assisting the LDCs

hours from the park. Observers of the experiment have reported that the enterprise appears to be slowly unraveling "the tie between education and work patterns, disrupting the transmission of beliefs to the young and undermining the ancestral view of how the world works and the proper place of human beings within that world."[74]

The traditional peoples of Amazonia have lived on the land for hundreds of years in settlements along the river. They live in communal houses called *malocas*, surrounded by crops that are planted in rotation. After several years, the *malocas* are moved to new locations. As a result, the people have flourished in an area with relatively poor soil and fragile rain forests. Recently, however, the Brazilian government has created permanent settlements that have begun to exhaust the land. The settlers threaten the land use practices of the indigenous population and now depend on imports of food and other resources.

In the Sierra Nevada de Santa Marta a similar pattern has emerged. For more than a thousand years, Indians lived in these coastal mountains. Communities ranged from coastal fishing villages to others high in the mountains, interconnected by a sophisticated system of roads and trading arrangements. The people carefully adapted their agricultural practices to the different conditions. As a result, some one hundred thousand people lived relatively prosperous lives. However, immigrants have moved into the area and tried to impose their own agricultural ideas. The fit has been a poor one; the immigrants' systems of production are not well suited to the land. Soil erosion and exhaustion, water shortages, and other environmental problems have diminished the population to twenty thousand who struggle to eke out an existence. The immigrants have rejected the highly integrated social and economic structure of the Indians as each family seeks to extract as much from the land as it can.[75]

Bolivia is one of the poorest countries in the Western Hemisphere; most of the poorest people live on the Altiplano, a high plateau between two ranges of the Andes Mountains. With its harsh, dry climate, the Altiplano is an extremely challenging place in which to make a living. Farmers grow potatoes, grains, and other crops on a plateau some 13,000 feet above sea level. Labor is plentiful on the Altiplano, but money and equipment are not.

One of Bolivia's greatest needs is to develop sources of clean water for drinking and irrigation. About one-half of all children born on the Altiplano die by the age of five; most deaths are a result of contaminated water. The Andean Children's Foundation (ACF) is one of hundreds of foundations working among the people of Bolivia. Its projects are designed by the local residents themselves. Staff members meet with villagers to plan projects and to ensure that projects are not just temporary but part of a long-term effort to improve the quality of life. Projects spark increased energy and efforts on the part of villagers to solve their problems. No attempt is made to "westernize" or "modernize" villages but rather to help reduce suffering, hunger, disease, and premature death.[76]

The ACF's hand-dug wells and hand-driven pumps are examples of small-scale, appropriate technologies. Local farmers purchase, install, operate, and maintain the pumps themselves. Farmers can borrow pumps and then build their own for about $25 in materials, or buy them for about $50. (Some models cost as much as $150.) One sign of success is that farmers with extremely modest resources were willing to buy the pumps themselves; one enterprising farmer was able to make his own pump from materials on hand and about five dollars worth of PVC pipe and cement after he watched his neighbor's pump operate. The pumps have been used to irrigate entire plots of land during dry spells as well as for greenhouses and family gardens.[77]

Grassroots development centers on the "empowerment of the poor, individually and collectively, *to acquire for themselves* and their children what they need and want for healthier, more fulfilling lives."[78] Women and children play central roles in formulating and implementing development plans. Development efforts seek to foster within villagers both the capacity to utilize (in sustainable ways) more of the local resources and benefits and the confidence to demand access to resources from those who control them. Both collective and individual empowerment are central. Empowerment involves "building awareness of alternatives, providing access to needed technologies and resources, building relevant skills, encouraging initiative and fostering local incentives, building effective organizations and community infrastructure, integrating development with existing cultural values and norms, and focusing on practical projects that have a high probability of successful completion."[79]

Appropriate technology—technology that is well integrated into local cultures and environments—is a key to successful development. Case studies in Latin America and elsewhere demonstrate the importance of these technologies in improving the quality of life for indigenous peoples, in helping to preserve the ecosystems on which they rely, and in preserving traditional cultural values. Such technologies are inexpensive and can be funded through microloan programs, which have generated great enthusiasm among development practitioners as well as villagers.[80] These technologies challenge traditional development efforts, however, and have been slow to generate support. Widespread dissatisfaction with development and foreign assistance policy may, however, open up the prospects for a grassroots-driven, appropriate-technology development effort that will foster local cultures and traditions.

Women and Development

Conventional development projects often do not benefit women. Economic growth does not necessarily translate into increased income for families because men often spend the extra income on themselves. Men grow cash crops, whereas women grow crops to feed their families. Development programs are aimed at men; development officials, almost exclusively men, generally limit their contact to men. Economic statistics focus on the value of goods exchanged; women's under- or unpaid contributions to the economy are not considered. As a result, development projects ignore women since the success of the projects is usually measured in terms of these statistics.

Women and children deserve particular priority and attention in development. The United Nations Children's Fund argues that development must begin with children: "A wise investment in children's health, nutrition and education is the foundation stone for all national development. Neglecting children's needs will, by contrast, condemn them and their society to a vicious circle of poverty and deprivation." UNICEF's agenda for reducing by the year 2000 infant and child mortality, maternal mortality, and severe malnutrition; increasing access to safe drinking water; and providing education for children and literacy programs for women can be accomplished for about $20 billion a year—2 percent of the world's annual military expenditures.[81]

Women play a critical role in grassroots development. A 1991 survey asked women from Africa, Asia, and Latin America what they most wanted and needed. They gave the following responses:

> Durable arrangements for the transfer of resources; reductions in (if not cancellations of) the debt burden; direct investment to meet capital requirements
>
> Equitable trading terms and better prices for primary commodities such as coffee, tea, and cocoa
>
> Access to credit and training; programs for awareness and confidence building
>
> Small to medium joint ventures to create jobs; continuing investments in sustainable economic growth
>
> Investments in the development and dissemination of appropriate technologies to reduce women's work burdens
>
> Access to good food, safe water, and education for both girls and boys
>
> Sustainable strategies for the use of natural resources
>
> Reallocation of financial resources to critical health care needs, including disease control, maternal and child health and family planning, and development of appropriate health systems

Options for Assisting the LDCs

Cooperation to establish, expand, and strengthen community-based approaches for promotional and educational activities of family planning and family life education

Access to information concerning women's bodies

The right to choose the number of children born and to plan families without government interference

Access to vaccines, medicines, and equipment

Universal access to contraceptives for both men and women.[82]

WOMEN AND COMMUNITY BANKS

One very valuable development activity is to expand women's access to small amounts of capital. Women in the developing countries often do not have access to capital for small-scale projects that require mills and other agricultural tools, sewing machines and material, and supplies to be used in commercial cooking ventures. These projects could help them improve their own and their families' quality of life and reduce their vulnerability to disease, drought, hunger, and other threats. The amount of money needed is not great; some loan programs provide loans of $50 to $100. Traditional development programs do not deal in such small amounts, nor do most commercial banks. Banks in many countries require women's husbands to sign for loans, and these men are rarely supportive of the idea. Nongovernmental organizations have begun to institute microloan programs in which women can borrow small amounts of money for family and commercial needs, and then repay the funds, with or without interest, when their income permits it. Funds are then made available to other women. The initial capital for the bank is usually donated by nongovernment organizations, which raise the money in developed nations.

The first microcredit program is generally believed to be the Grameen Bank, started by Muhammad Yunus, a Bangladeshi economist who lent the equivalent of $26 to forty-two workers in

1976. With the 62 cents per person, the recipients bought materials for the products they made, sold the finished goods, and repaid the debt. Two million people have borrowed money from Grameen banks; one-third of them have increased their income to above the poverty level and another third are close to doing so. The idea has expanded to 8 million people in forty-three countries including economically depressed areas in the United States, because of the bank's ability to help poor people, create jobs, and build confidence and self-sufficiency. But microlending only receives two percent of the $60 billion in annual spending on international development. A February 1997 meeting in Washington, D.C., by proponents of microlending committed to expand the practice to 100 million people by 2005.[83]

The growth of community banks has been phenomenal during he late 1980s and early 1990s. Despite their recent vintage, these banks have already shown themselves to be one of the most promising approaches to development ever introduced. NGOs have escaped many of the problems that have plagued centralized government development agencies, and their efficiency is widely recognized. Similarly, grassroots development programs are more likely to address the needs and concerns of communities, families, and women than are traditional programs. NGOs and grass-roots development groups have been the leading champions of community banks; these banks can be part of a comprehensive development effort aimed at helping women and their families increase their food production, health, literacy, and overall quality of life. The funds dedicated to community banking are dwarfed by traditional development assistance programs; much more money could be used to fund new banks and reach more women and families throughout the world.

Community banks are known by a variety of names—development banks, microenterprise credit projects or programs, microlending banks or institutions, women's development banks, poor people's banks, peer lending groups, and so forth. In the United States, community banks have sprung up in rural and urban areas throughout the country, from east Oakland and south Chicago to small towns in Vermont to the Pine Ridge Indian Reservation in South Dakota. Some organizations loan money directly to residents; others involve traditional banks. In San Francisco, the Women's Initiative for Self-Employment provides loans and assistance to poor and homeless women and recent immigrants. Some community banks take a holistic approach to serving a small number of women. Services include skills training, credit, marketing assistance, basic education, family planning, and health and nutritional services.[84] Area programs offer comprehensive services in specific economic areas. They address specific problems of microenterprises and self-employment and become active in marketing, input supply, technology, training, and credit.

Toward Sustainable Development

The Bruntland Commission defined the global task as "development that meets the needs of the present without compromising the ability of future generations to meet their own needs." The Commission also emphasized that this is not just a task for the LDCs: "'Sustainable' development becomes a goal not only for the 'developing' nations but for the industrial ones as well." But we know little about what kinds of actions are sustainable: can we continue to rely heavily on the burning of fossil fuels? Since supplies are limited and the environmental consequences serious, fossil fuel use is, in the long run, unsustainable. But for how long and at what levels can we continue to use these fuels until we are ready to move to alternative fuels? As with the debate over the carrying capacity of land, it is difficult to determine what is sustainable since ecological conditions and technologies that affect pollution levels are constantly in flux.[85]

Two efforts are critical in bridging the gap between MDC responsibility for most environmental problems and LDC susceptibility to the consequences of these problems: First, leaders of both the North and the South must be convinced that their interests and survivability are inextricably linked; second, technologies and resources must be transferred from the North to the South.[86] The discussion above focuses on incremental steps that could be done to increase assistance to the South. But what are the prospects for convincing policy makers in the North and the South to accept that link? As developing countries modernize their economies, they can do so while protecting the environment. In the long run, building economies in environmentally sound ways is also economically viable, but it will require enormous initial expenditures and the most advanced technologies. Although supplying funds is important, developed countries can make available the most advanced technologies for efficient and environmentally benign energy production to other countries. Unfortunately, developed nations have traditionally seen the developing world primarily as a source of cheap raw materials and a dumping ground for obsolete technologies. That view has greatly exacerbated global environmental decline.[87] Linking environmental protection with development is of great interest to both the North and the South. The South can get the assistance it needs for ecologically sustainable development while the North gets support for the environmental goals it seeks. As we rethink and revise development efforts, both the North and South and current and future generations stand to benefit.

It is also possible that the kinds of incremental policy steps discussed above will not be sufficient to prevent global ecological catastrophe. Nor will they be effective enough to remedy global inequality. From an ecological science perspective, what level of environmental quality is necessary to sustain and perpetuate life? From a public philosophy perspective, at what level of economic activity and consumption, and at what level of inequality do we want to sustain life? From a political perspective, what level of change is possible—how much commitment to and capacity for change is there? Even as we struggle with incremental policy steps to link the futures of the North and South more closely together, we need to pursue the prospect that more fundamental changes will be required. We must recognize the importance of linking improvements in the quality of life in the less-developed world with a more ecologically sustainable global political economy.

References

1. Michael E. Moseley, *The Incas and Their Ancestors: The Archaeology of Peru* (London: Thames and Hudson, 1992).
2. See, generally, Energy and Environmental Study Institute, *Partnership for Sustainable Development* (Washington, D.C.: EESI, 1991); Jessica Tuchman Matthews, ed., *Preserving the Global Environment: The Challenge of Shared Leadership* (New York: Norton, 1991); James J. MacKenzie, *Breathing Easier: Taking Action on Climate Change, Air Pollution, and Energy Insecurity* (Washington, D.C.: World Resources Institute, 1989); Robert Repetto, *World Enough and Time: Successful Strategies for Resource Management* (New Haven, Conn.: Yale University Press, 1986); World Commission on Environment and Development, *Our Common Future* (Oxford: Oxford University Press, 1987); Oran Young, *International Cooperation: Building Regimes for Natural Resources and the Environment* (Ithaca, N.Y.: Cornell University Press, 1989); Peter H. Sand, *Lessons Learned in Global Environmental Governance* (Washington, D.C.: World Resources Institute, 1990); Richard Elliot Benedick, *Ozone Diplomacy* (Cambridge, Mass.: Harvard University Press, 1991); E. D. Brown and R. R. Churchill, eds., *The UN Convention on the Law of the Sea: Impact and Implementation* (Honolulu: Law of the Sea Institute, University of Hawaii, 1985); Jacques G. Richardson, *Managing the Ocean: Resources, Research, Law* (Mt. Airy, Md.: Lomond Publications, 1985); Lynton Keith Caldwell, *International Environmental Policy: Emergence and Dimension* (Durham, N. C. and London: Duke University Press, 1990); World Resources Institute, *The Crucial Decade: the 1990s and the Global Environmental Challenge* (Washington, D.C.: World Resources Institute, 1989); and World Resources Institute, *World Resources, 1990-91* (Washington, D.C.: WRI, 1990).
3. Christopher Flavin, "Slowing Global Warming: A Worldwide Strategy," Worldwatch paper 91 (Washington, D.C.: Worldwatch Institute, 1989), 26.
4. World Resources Institute and International Institute for Environment and Development, *World Resources 1988-1989* (New York: Basic Books, 1988), 246-47, 306-07.
5. See, generally, Norman Myers, *The Primary Source* (New York: Norton, 1984); and E. O. Wilson, ed., *Biodiversity* (Washington, D.C.: National Academy Press, 1988).
6. Commission on Developing Countries and Global Change, *For Earth's Sake* (Ottawa: International Development Research Centre, 1992), 31-32.

7. Population Reference Bureau, "1989 World Population Data Sheet" (Washington, D.C.: PRB, 1989).
8. Commission on Developing Countries and Global Change, *For Earth's Sake*, 23.
9. Quoted in Mark Hertsgaard, "The View from 'El Centro del Mundo,'" *Amicus Journal* (Fall 1992): 13.
10. Commission on Developing Countries and Global Change, *For Earth's Sake*, 41–42.
11. See Benedick, *Ozone Diplomacy*; Seth Cagin and Philip Dray, *Between Earth and Sky* (New York: Pantheon, 1993); and Vandana Shiva, "The Greening of the Global Reach" in Wolfgang Sachs, *Global Ecology: A New Arena of Global Conflict* (London: Zed Books, 1993).
12. United Nations Development Program, *Human Development Report 1991* (New York: Oxford University Press, 1991), 166.
13. James Gustave Speth, "The Rich Are Getting Richer While the Poor Waste Away," reprinted in the *Deseret News*, 25 August 1996, A20.
14. Alan B. Durning, *Poverty and the Environment: Reversing the Downward Spiral*, Worldwatch Paper 92 (Washington, D.C.: Worldwatch Institute, 1989), 20.
15. UN Development Program, *Human Development Report*, (New York: Oxford University Press, 1996).
16. World Resources Institute, *World Resources 1990–91* (Washington, D.C.: World Resources Institute, 1990), 256–64.
17. Jodi L. Jacobson, "The Forgotten Resource," *World Watch* (May–June 1988): 35, 36–38.
18. Ibid., 38–39.
19. Ibid., 36.
20. Ann Misch, "Lost in the Shadow Economy," *World Watch* (April 1992): 18–19.
21. Ibid., 20–21.
22. Ibid., 25.
23. Morris Miller, *Debt and the Environment: Converging Crises*, (New York: United Nations Publication, 1991), 46.
24. Andy Pasztor, "White House Girds to Promote Huge Arms Sales to Many Nations," *Wall Street Journal*, 24 July 1992; Robert Pear, "U.S. Ranked No. 1 in Weapons Sales," *New York Times*, 11 August 1991.
25. *World Watch* (January–February 1995): 39.
26. Jane Perlez, "African Dilemma: Food Aid May Prolong War and Famine," *New York Times*, 12 May 1991.
27. Sylvia Nasar, "Third World Embracing Reforms To Encourage Economic Growth," *New York Times*, 8 July 1991.

References

28. The World Bank, *World Development Report 1988* (New York: Oxford University Press, 1988).
29. Jyoti Shankar Singh, "How to Reduce the Debt Burden," *Earthtimes* (1–15 October 1996), 12.
30. Paul Lewis, "Debt-Relief Cost for the Poorest Nations," *New York Times*, 10 June 1996, C2; Paul Lewis, "I.M.F. and World Bank Clear Debt Relief," *New York Times*, 30 September 1996, C2; editorial, "The Third World Debt Crisis," *New York Times*, 26 June 1996, A14.
31. Gary Gardner, "Third World Debt Is Still Growing," *World Watch* (January–February 1995): 37–38.
32. Alan B. Durning, *Poverty and the Environment: Reversing the Downward Spiral*, Worldwatch Paper 92 (Washington, D.C.: Worldwatch Institute, 1989), 40–41.
33. This material is compiled from Lester Brown et al., *The State of the World 1990, 1991,* and *1992* (Washington, D.C.: Worldwatch Institute, 1990, 91, 92); and Global Tomorrow Coalition, *The Global Ecology Handbook* (Boston: Beacon Press, 1990), chs. 1, 4, and 5.
34. "Global Food Supplies: Tighter but Adequate," *Wall Street Journal*, 15 July 1996, A1.
35. World Commission on Environment and Development, *Our Common Future* (New York: Oxford University Press, 1987).
36. "The Tragedy of the Commons," *Science* 162 (1968): 143.
37. Thomas F. Homer-Dixon, Jeffrey H. Boutwell, and George W. Rathjens, "Environmental Change and Violent Conflict," *Scientific American* (February, 1993): 38–45.
38. Gordon J. MacDonald, "Technology Transfer: The Climate Change Challenge," *Journal of Environment and Development*, 1 (Summer 1992): 1–40.
39. Amory Lovins and Hunter Lovins, "The Next Energy Crisis? Efficient Energy Use vs. Producing More Power," *Popular Science* 249 (September 1996): 89.
40. The Environmental Technology Transfer Board, established by the U.S. EPA in 1989, has focused on these issues. For a discussion of its work, see MacDonald, "Technology Transfer," 11–12.
41. Ruth Greenspan Bell, "EPA's International Assistance Efforts: Developing Effective Environmental Institutions and Partners," *Environmental Law Review* 24 (October 1994): 10593–99.
42. S.C. Majumdar, P. N. Bhattacharya, and S. T. Rao, "Turning Brain Drain into Brain Gain To Solve Environmental Problems," *Environmental Manager* 1 (June 1995): 16–18.

43. A. O. Adede, "International Environmental Law from Stockholm to Rio—An Overview of Past Lessons and Future Challenges," *Environmental Policy and Law* (22/2 1982): 100–101.
44. For an introduction to this literature, see Walter Rosenbaum, *Environmental Politics and Policy* (Washington, D.C.: Congressional Quarterly Press, 1993).
45. Robert Repetto, "Accounting for Environmental Assets," *Scientific American* (June 1994): 94–100, at 94.
46. Ibid., 96.
47. Commission on Developing Countries and Global Change, *For Earth's Sake*, 22–23.
48. Peter Truell, "The Outlook: Debt Relief Endangers Third World Prospects," *Wall Street Journal*, 6 May 1991.
49. See Smithsonian Institution and the Natural Resources Defense Council, "Debt-for-Nature Swaps: Progress and Prospects (Washington, D.C.: Natural Resources Defense Council, 1991).
50. Gary Gardner, "Third World Debt Is Still Growing," *World Watch* (January–February 1995): 37–38, at 38.
51. Miller, *Debt and the Environment*.
52. See Richard Elliot Benedick, *Ozone Diplomacy*.
53. William Stevens, "Talk Seeks to Prevent Huge Loss of Species" *New York Times*, 3 March 1992.
54. "What's New at the Bank?" *Environment Matters* (Summer 1996): 26; "World Bank Environmental Projects, July 1995–June 1996" (Washington, D.C.: World Bank, 1996).
55. Lee A. Kimball, *Forging International Agreement* (Washington, D.C.: World Resources Institute, 1992), 73.
56. Ibid., 73.
57. Ibid., 74.
58. World Resources Institute, *World Resources 1994–95* (Washington, D.C.: WRI, 1994), 231.
59. Ibid., 227.
60. Ibid., 229.
61. The World Bank, *Monitoring Environmental Progress: A Report on Work in Progress* (Washington, D.C.: World Bank, 1995), 57–64.
62. The World Bank's conference, "Rural Well-Being: From Vision to Action," was held in September 1996; proceedings of the conference will likely be available from the Bank by the end of 1997.
63. World Resources Institute, *World Resources 1994–95*, 232.
64. United Nations Development Report, *Human Development Report 1991* (Oxford: Oxford University Press, 1991), 14.
65. Gabriel Almond and G. Bingham Powell, *Comparative Politics: System, Process, and Policy* (Boston: Little, Brown, 1978), 252.

References

66. Ibid., 72.
67. For a critical review of the traditional approach to development, see Kate Manzo, "Modernist Discourse and the Crisis of Development Theory," *Studies in Comparative International Development* 26, no. 2 (Summer 1991).
68. See, for example, David Apter, *Choice and the Politics of Allocation* (New Haven, Conn.: Yale University Press, 1971).
69. One of the earliest theorists to discuss dependency theory was Paul Baran, *The Political Economy of Growth* (New York: Monthly Review Press, 1957).
70. Charles D. Kleymeyer, "Cultural Energy and Grassroots Development," *Grassroots Development* 16:1 (1992): 22–31.
71. Timothy S. Evans, "Building an Appropriate Development Model: Preliminary Foundations," unpublished manuscript (April 1991), 12.
72. Kleymeyer, "Cultural Energy and Grassroots Development," 28.
73. William Ophuls and A. Stephen Boyan, Jr., *Ecology and the Politics of Scarcity Revisited* (New York: W. H. Freeman, 1992), 176–77.
74. Kleymeyer, "Cultural Energy and Grassroots Development," 24–25.
75. Commission on Developing Countries and Social Change, *For Earth's Sake*, 46–47.
76. Timothy S. Evans, "Important Information about CHOICE" (n.d.)
77. Timothy S. Evans, "Hand-Dug Wells, Hand-Driven Pumps: Self-Help Water Resource Management in Rural Andean Communities, Summary of Progress to Date, Handpump Investigation" (14 September 1991).
78. Timothy S. Evans, "Building an Appropriate Development Model: Preliminary Foundations" (April 11, 1991), 4.
79. Ibid., 5–8.
80. See, for example, Donald R. Katz, "Where Credit is Due," *Investment Vision* (August–September 1991): 48–57.
81. United Nations Children's Fund, *The State Of The World's Children* (New York: United Nations Publications, 1990), 66.
82. "What do Women Want?" panel discussion held at "Women's Health: The Action Agenda for the Nineties," 18th Annual National Council on International Health Conference, Arlington, Va., 23–36 June 1991, reported in Jodi L. Jacobson, "Closing the Gender Gap in Development," in Lester Brown et al., *State of the World 1993* (New York: Norton, 1993), 77.
83. Editorial, "Micro-Loans for the Very Poor," *New York Times*, 16 February 1997, E12; the Microcredit Summit was organized by RESULTS Educational Fund, microcredit@igc.apc.org.
84. Katharine McKee, "Microlevel Strategies for Supporting Livelihoods, Employment, and Income Generation of Poor Women in the Third

World: The Challenge of Significance," *World Development* 17 (1989): 993, 995.
85. Nathan Keyfitz, "The Right Steps at the Right Time," *Ecodecision* (June 1993): 31.
86. For more in this argument, see Marnie Stetson, "People Who Live in Green Houses . . ." *World Watch* 4, no. 5 (September–October 1991): 22–29.
87. For a general discussion of development and environmental protection, see Repetto, *World Enough and Time*.

7

Prospects for Preserving the Global Environment

ONE OF THE great ironies of modern life is that as we come to realize the consequences of our Western model of consumption, industrialization, and hostility to nature—all of which are at the heart of the environmental crisis—we have begun to recognize that we have much to learn from those who have yet to embrace all the trappings of modernity. The authors of Canada's National Report, which was prepared for the United Nations Conference on Environment and Development, recognized the importance of the wisdom of native peoples:

> As people who have lived in harmony with nature and close to the land for centuries, aboriginal peoples of Canada have developed an immensely valuable information base and expertise which can be shared with the rest of Canadian society. As long-standing custodians of this traditional knowledge, they can provide detailed information on the workings of natural ecosystems, and can provide a perspective and information on how environmental systems function over time.[1]

Indigenous peoples have profound knowledge of how ecosystems operate and what is required to sustain them.[2] However, we have done much over the past five hundred years to eliminate the knowledge that we now find so compelling. The Spanish invasion of the New World set the tone for what was to follow. The quest for wealth permeated the European settlement of the western hemisphere. The relentless westward expansion in the United States and efforts to exterminate Native Americans in the eighteenth and nineteenth centuries were a continuation of the conquest that began in 1492, despite attempts at resistance. The European view, which was at the heart of the conquest and which continues today, is rooted in a kind of spiritual blindness. The conquistadors were blinded by their consuming greed; they could not appreciate the cultural, ecological, and spiritual values practiced in the New World. We have now become so enamored of technological progress that we often fail to recognize and understand enduring truths that come from what appear to be unsophisticated sources.

Some believe that we have developed the mechanical and technological power to dominate and control nature, that traditional or indigenous knowledge is unnecessary. They assume that we can indulge ourselves now and rely on some technological fix to save us in the future.[3] Nevertheless, it is increasingly clear that we are just as dependent on nature as were the early inhabitants of these lands. We can escape this environmental accounting in the short run; we can avoid the costs resulting from environmental degradation for a time. But we will eventually have to face the consequences of our actions. How should we respond to the great uncertainties surrounding the causes and consequences of global environmental concerns? How can we make incremental change, learn from experience and make the requisite adjustments, and still take action that is sufficient to avoid catastrophes? How much support is there for developing new ways of thinking about how we interact with nature? What kinds of ideas might serve as the basis for a ecologically sustainable economic life?

The Challenge of Scientific Uncertainty

Great uncertainty surrounds the causes and consequences of pollution; and policy makers form decisions within this uncertainty. In

The Challenge of Scientific Uncertainty

many cases, adverse health effects and other negative consequences may not become apparent for many years. Sometimes the effects of environmental hazards are are largely irreversible, either in terms of loss of human life or ecological changes. Policy makers must include learning from experience and adjusting.

There is little agreement over how much we need to know about the health and environmental effects of pollutants before taking regulatory action, what the balance should be between waiting for more research and using policy interventions to prevent harm, and how much risk from environmental pollution we should accept. A major issue is how risks should be calculated: Some argue that all persons should be protected, including those who are most susceptible to the effects of pollution. Others insist that the risk posed to members of the community in general or on average should be the basis of regulatory action. Once we estimate risks, a second round of questions grapple with how we should balance reducing environmental risks with other values, such as individual freedom and corporate autonomy.

In some ways, the state of scientific understanding concerning the effects of pollution is much like the research on the health effects of smoking in the 1950s and 1960s. As evidence mounted during those decades on the risks of tobacco use, the tobacco industry challenged the studies as being inaccurate. Following that pattern, industry groups continue to challenge the idea that industrial emissions pose public health threats. Given the long latency period between exposure to some pollutants and the onset of disease, the difficulties in tracing the source of pollutants that have done damage, and other characteristics of the interaction of pollution and public health, many have demanded that more research be done before control measures are imposed.[4] Epidemiological studies are particularly useful since they can help identify long-term results from exposure to pollutants. However, since these studies cannot *prove* that a particular pollutant causes specific health problems, they are often the target of criticism by industry officials.

Scientific knowledge about environmental problems is growing, but it is still minimal when compared with what scientists and policy makers would like to know. The range of potential policy responses is immense. For some concerns, such as the destruction of the stratospheric ozone layer, scientific agreement concerning the need for action is widespread. For other concerns such as global climate change, unambiguous evidence of warming may not be available until it actually begins, but by then it may be irreversible.

Studies have emphasized the importance of separating risk assessment from risk management—the making of policy choices in response to these assessments. The *assessment* of risks in environmental and health regulation, for example, includes laboratory and field observation of adverse health effects and exposure to particular agents; information on extrapolation methods for high to low dose and animal to human; field measurements, estimated exposure, and characterization of populations; hazard identification (does the agent cause the adverse effect?); dose-response assessment (the relationship between dose and incidence of problems in humans); exposure assessment (what exposures are currently experienced or anticipated under different conditions); and risk characterization (the estimated incidence of the adverse effect in a given population). In contrast, the *management* of risk includes developing regulatory options; evaluating public health and the economic, social, and political consequences of regulatory options; and determining what actions to take.[5]

Although the separation between scientific assessment and the analysis of policy options is difficult if not impossible to maintain, it is particularly important to recognize this separation when making policy in areas of great scientific uncertainty. Policy priorities and commitments can easily color perceptions and interpretations of scientific research, particularly with no consensus on the scientific issues.

The management of risk poses at least two dilemmas. Policy makers often have to choose among the arguments of competing scientists and experts. This is particularly difficult for nonexperts. There are a number of ways that policy makers might deal with the uncertainty stemming from divergent scientific views—from traditional adversarial proceedings to science courts, where scientists replace lawyers in judicial-style hearings. More formal mech-

The Challenge of Scientific Uncertainty

anisms have been proposed to bring contending scientists together to question one another and to be questioned by policy makers in an effort to help sort out the implications of scientific disagreements and to identify areas of agreement and disagreement. Developing an effective process for addressing these concerns is critical in devising appropriate policies.[6]

Further, unless policy making authority is to rest in the hands of a technocratic elite or body of "experts," scientific research and policy choices must be shared with the public. How this sharing takes place has important implications for future policy choices. To be effective, public policy debate requires clear statements of risks and the identification of costs and benefits, but these requirements do not mesh well with the slowness, deliberativeness, and uncertainty of the scientific process.[7]

Policy makers need to communicate the nature of hazards to the public in a way that is dramatic enough to get people's notice, but not so dramatic as to invoke demands for rash or inappropriate action. The risks associated with global environmental problems are sufficiently frightening that they have captured widespread attention. But careful dissection of the scientific and technical issues involved can put a great strain on the democratic processes. It is not easy to determine the optimal level of detail and technical specifics to inform public policy debates.[8]

In dealing with uncertainty, policy makers might pursue at least two kinds of strategies, beyond simply prohibiting every action for which there is a projected or hypothesized risk. The first approach is to do nothing and leave to private institutions and individuals the responsibility for responding to environmental challenges. Such a position may be the result of a number of different concerns and motivations. Policy makers may opt to wait until research findings are sufficiently clear to compel action, but such a position can also be nothing more than a stalling technique by a regime unwilling to impose new costs on their constituents or a way to escape responsibility for difficult or unpopular actions. A more responsible position is to argue that policy interventions can be more efficient and effective if additional research is completed before policies are designed.[9]

A more sophisticated justification for taking no action asserts that regulatory interventions designed to "play it safe" by seeking to anticipate and prevent uncertain environmental and other risks are more dangerous than coping with problems as they develop. Aaron Wildavsky has argued that the costs of intervening to prevent possible damage are not always well recognized or appreciated. Resources are exhausted "in a futile effort to anticipate the future"; these resources would otherwise be available to respond to problems once they become evident, to mitigate effects once they occur. Society should foster resilience rather than trying to anticipate potential danger. Resilience does not mean simply doing nothing and waiting for a problem to occur; it requires "preparing for the inevitable . . . by expanding general knowledge, technical facility, and command over resources," rather than directing scarce resources toward preventing specific but hypothesized harms.[10]

A second strategy is to take preventive actions whenever either the probability of a harm occurring is quite high, or when the probability of an event occurring is low but the risks associated with its occurrence are so great as to justify anticipatory action. The traditional idea of insurance provides some conceptual guidance here. Environmental threats are no different from other challenges that confront human beings. It is prudent to insure against possible adverse changes, particularly against those whose consequences might be serious or even irreparable. The kinds of changes associated with a rapid rise in the average temperature of the earth, some have argued, are simply so cataclysmic that the risk of such developments must be minimized. Similarly, construction safety codes for highways, bridges, and buildings require architects and construction companies to prepare for contingencies such as earthquakes. These precautions raise construction costs considerably, but are viewed as prudent.

To summarize, in situations in which uncertainty is great but the consequences of errors are likely to be catastrophic, policy makers can (1) set priorities to ensure that the most critical uncertainties receive the most attention, (2) take actions to protect against the worst consequences of error, and (3) establish mechanisms to

The Challenge of Scientific Uncertainty

monitor phenomena and research efforts so that we learn from experience.[11]

Scientific Uncertainty and Policy Action

The uncertainties surrounding global environmental issues are at the heart of both national and international policy debate. For some, uncertainty will be an excuse for some not to take action. Uncertainty will prod others to insure against potentially catastrophic events. Both choices will have consequences. Uncertainties ultimately require attention to ensuring that policy making is a dynamic, flexible process that includes regular learning and adaptation as changes in the world unfold.

In many cases, regulatory actions taken to insure against one hazard may also mediate against other harms. Reducing emissions of chlorofluorocarbons might reduce the threat of global warming; such a reduction will also very likely reduce the loss of the stratospheric ozone layer. Steps to increase energy efficiency will lead to lower energy prices and, consequently, to lower prices for manufactured goods and less dependence on imported oil; reduced health hazards from local air pollution; diminished acid rain emissions; and reduced carbon dioxide emissions. Investments that can yield multiple benefits are clearly worthwhile. Even with uncertainty, however, the question may not be whether to buy insurance (or invest in actions to reduce the likelihood or magnitude of change), but how much insurance to purchase.[12]

The precautionary principle, widely cited in international law and environmental discussions, may provide some guidance in dealing with uncertainty. The principle requires that we err on the side of safety or precaution: Rather than letting market forces determine actions, government regulation sets limits on what possibilities for harm are permitted. But the precautionary principle is incomplete; it does not tell us what are the greatest risks, or how much precaution we should take.[13]

In formulating global environmental policy, we can try to reduce as much risk as possible, given the resources available. Since the number of environmental and health hazards is much greater than can be regulated, government resources must be carefully rationed to ensure that the most serious risks are addressed first and that hazards will not simply be transferred from one medium to another. In theory, policy makers should rank risks and set priorities. In the United States, however, the EPA is limited by statutes that do not treat environmental problems in such a unified way. Different statutes authorize the agency to pursue different problems. These laws take different approaches to reducing risk, from balancing risks against the benefits and costs of regulation to prohibiting cost considerations and permitting only considerations of risk and health. The EPA must implement laws, meet congressionally and judicially imposed deadlines, and respond to other demands for regulatory action.[14]

Rational, effective policy making assumes that different risks can be compared. But risk comparison is often extremely difficult. Ocean dumping, for example, may reduce human exposure to toxic wastes but increase damage to sea life. Nor do we always choose to save the maximum number of lives—we demand that air transportation be safer than driving, despite the much greater loss of life in motor vehicle accidents. Regulatory agencies must also respond to public fears and concerns. Regulation of different substances may be constrained by different political factors even though the level of risk may be quite similar. Most people tend to overestimate the frequency and seriousness of "dramatic, sensational, dreaded, well-publicized causes of death and underestimate the risks from more familiar, accepted causes that claim lives one by one."[15]

Social psychologists argue that voluntary risks are more acceptable than involuntary ones; risky situations in which individuals have some measure of control are less feared than those in which individuals do not; risks that are perceived to be unfairly distributed are more unpopular than those that are not; familiar risks are less feared than unfamiliar ones; hazards that are concentrated or dramatic and result in many deaths are more feared than those whose victims are widely dispersed; and some illnesses and diseases are more dreaded than others. This does not mean that the public

The Challenge of Scientific Uncertainty

is wrong or uninformed or that these views diverge from the "true" nature of risk, but rather that risk is a complicated concept.[16] It includes not only the calculated probabilities of death or illness but also their interaction with several other socially important values.[17] Policy makers have few guidelines to help them establish priorities for action. Much more important than comprehensive, rational attempts to compare risks are the vagaries of politics.

This discussion is infused with values: What risks are we willing to take? What do we want most? What kind of world do we want to leave to our children? What risks are we willing to leave to our children? Other issues are implicated. How, for example, do we balance obligations to future generations with obligations to current generations? Do we fail to promote development in the Third World now because to do otherwise would make the world less hospitable to our children? Is it fair to give priority to the unborn? Is this trade-off necessary or inevitable? Can economic growth be sustainable? At what price? What if it requires sacrifices of the current generation? Many argue for the advantages of a less materialistic, consumption-oriented life-style. One can even argue that we do not have the right to consume more than is necessary to sustain ourselves while others lack basic necessities, an obligation that is just as binding in terms of those who currently share the planet with us as those who come after us.

Perhaps it is easier to argue that we do not have the right to enjoy the benefits of certain activities while we impose the consequences of our actions on others. For example, we enjoy the benefits of nuclear power now, but we pass along the burden of caring for the waste to future generations. One could answer that if the resultant power produced some benefits that could be passed on to future generations, then it would be fair to impose some of the burden on them. A prudent response would be to protect against the worst-case scenario. This is a simple notion of fairness as a form of equal opportunity (equal access to common, scarce resources) and as a way to avoid externalities.

We are coming to recognize the "ineffable and subtle intertwining of living organisms on the Earth," the interdependence of life.[18]

Microbiologists have taught us the stunning fact that virtually all forms of life are composed of DNA—that humans are composed of essentially the same subunits as are animals and plants, that our differences are simply differences in the arrangements of subunits. Our world is fragile, our rapacity can harm it irreparably.

Environmental protection is fundamentally a moral issue. Air and water pollution that threatens public health also violates basic norms of justice: Those who profit from pollution-producing activities are often not the ones who live downwind or downstream. Throughout the world, the wealthy and the powerful consume resources and produce pollution and waste at levels hundreds of times greater than those living in the poorest nations. The poor and the powerless suffer the most from global environmental problems, yet they contribute the least to them. They are the first to suffer, but have the fewest material and political resources to protect themselves. Generations yet unborn will bear the burdens of guarding our nuclear wastes and cleaning up our polluted lakes and rivers.[19]

An environmental ethic of stewardship and protection of natural resources is justifiable from many points of view. A simple calculation of our common self-interest should prompt us to take the steps required to secure a sustainable environment for ourselves and our children. Requiring industries, households, and other institutions to bear the full—environmental and natural-resource—costs of their activity is a basic element of the fair operation of markets.

Such optimists as Gregg Easterbrook believe that pollution in the Western world "will end within our lifetime," that the "most feared environmental catastrophes, such as runaway global warming, are quite unlikely" that environmentalism, "which binds nations to a common concern, will be the best thing that's ever happened to international relations," and that "nearly all technical trends are toward new devices and modes of production that are more efficient, use fewer resources, produce less waste, and cause less ecological disruption than technology of the past."[20] Others find that "just about every important measure of human welfare shows improvement over the decades and centuries"—life expectancy, price of raw materials, price of food, cleanliness of the environ-

The Challenge of Scientific Uncertainty

ment, population growth, survival of species, and the quantity of farmland.[21] Tom Tietenberg and other economists argue that we will never run out resources—the earth's air, water, and crust will serve us for millions of years to come. The problem is not the existence of these resources, "but whether we are willing to pay the price to extract and use those resources." However, some resources will become increasingly scarce in some localities, and rise in extraction costs will cause consumption of some resources to decline. The failure of markets to incorporate environmental costs, such as pollution, climate modification, and loss of genetic diversity produces "falsely optimistic signals and the market makes choices that put society inefficiently at risk."[22]

The optimists may be right in claiming that human ingenuity can respond to these problems and reverse these troubling trends. The trends in environmental damage, resource loss, and food production are all negative, but they may be reversible. Such changes, however, pose a tremendous challenge to our ingenuity, our governments, and our collective and individual wills. "We may be smart enough to devise environmentally friendly solutions to scarcity," one scholar has written, but we must emphasize "early detection and prevention of scarcity, not adaptation to it." But if we are not as smart and as proactive as optimists claim we are, "we will have burned our bridges: the soils, waters, and forests will be irreversibly damaged, and our societies, especially the poorest ones, will be so riven with discord that even heroic efforts at social renovation will fail."[23] Conservation of lands—setting aside areas for protection or nondevelopment—is not enough; human activity is pervasive, and opportunities for establishing wilderness, refuge, and other protected areas are very limited. Humans are an integral part of the earth's ecology; the primary challenge is to find ways of meeting human needs while protecting natural resources.[24]

The debate between optimists and pessimists over the state of the planet cannot be easily resolved. The scientific evidence that the earth faces extremely serious environmental threats is, for me, quite compelling. The more important debate may be over the assumptions and values that frame these scientific and policy disputes.

Since global environmental threats have the potential to do tremendous harm, and the distribution of those harms largely fall on people (those of this generation who live in poverty and future generations) who benefit least from the actions that produced the threats, precautionary actions are obligatory.

Public Support for Environmental Protection

One barrier to effective implementation of global environmental accords is that the costs of taking action now to prevent future problems will be immediate, but the benefits, in terms of less pollution or more resources or a more sustainable biosphere, will not accrue until far into the future. Future generations have no political power now. Short-run electoral incentives have a firm grasp on every politician who wishes to continue his or her career. How can we generate the political support to impose costs on us now, while there is no immediate, cathartic crisis that looms over us, but only the probability that we will, by acting now, save future generations from grief?

Despite the common belief that only people in wealthy nations are concerned about environmental quality, a recent poll by the Gallup International Institute found that people in both wealthy and poor nations express strong support for environmental protection; "majorities in most of the 24 nations surveyed by Gallup international affiliates give environmental protection higher priority than economic growth."[25]

Table 7.1 shows tremendous differences in views among people in the former Soviet Union and the industrialized democracies concerning environmental quality, reflecting the neglect of air and water pollution and hazardous wastes in the former communist countries. Table 7.2 demonstrates the immediacy of environmental problems, such as polluted water and lack of sanitation, in the less-developed world. Table 7.3 compares how people in different parts of the world view the state of the environment. Most people seem to realize the seriousness of these problems.

Public Support for Environmental Protection

TABLE 7.1

RATINGS OF ENVIRONMENTAL QUALITY AT LOCAL, NATIONAL, AND WORLD LEVELS*
Percentage saying "very/fairly bad"

	In community	In nation	In world
Poland	71	88	73
Russia	69	88	66
Korea (Rep.)	57	74	66
Hungary	48	72	71
India	44	52	42
Turkey	44	42	45
Chile	41	68	88
Brazil	41	49	64
Nigeria	34	38	24
Mexico	31	56	70
Japan	31	52	73
Portugal	30	39	75
Philippines	28	52	58
United States	28	45	66
Uruguay	28	37	74
Great Britain	27	36	76
Netherlands	24	45	84
Germany (West)	22	42	86
Switzerland	20	27	86
Canada	18	26	79
Finland	13	13	73
Denmark	12	18	92
Ireland	10	14	73
Norway	10	12	88

*Respondents were asked: "Now let's turn our discussion to the environment. When we say environment, we mean your surroundings—both the natural environment—the air, water, land and plants and animals-as well as buildings, streets and the like. Overall, how would you rate the quality of the environment in our nation—very good, fairly good, fairly bad, or very bad? In your local community? In the world as a whole?"

SOURCE: Riley E. Dunlap, George H. Gallup, Jr., and Alec M. Gallup, *Health of the Planet* (Princeton, N.J.: George H. Gallup International Institute, 1993).

TABLE 7.2

PERCEIVED SERIOUSNESS OF ENVIRONMENTAL PROBLEMS IN LOCAL COMMUNITY*
Percentage saying "very serious"

	Poor water	Poor air	Contaminated soil	Inadequate sewage	Too many people	Too much noise
North America						
Canada	17	17	11	21	7	5
United States	22	18	12	17	11	7
Latin America						
Brazil	43	30	24	49	21	24
Chile	13	18	17	33	9	17
Mexico	25	21	24	39	23	23
Uruguay	7	9	9	25	7	10
East Asia						
Japan	14	12	9	15	5	8
Korea (Rep.)	35	32	14	40	18	28
Philippines	23	12	11	30	9	13
Other Asia						
India	49	53	19	46	45	50
Turkey	42	37	27	62	38	29
Eastern Europe						
Hungary	15	19	12	17	5	9
Poland	62	61	50	55	10	24
Russia	39	39	28	40	10	12
Scandinavia						
Denmark	5	4	4	3	1	3
Finland	17	19	14	12	4	4
Norway	13	13	10	11	8	9
Other Europe						
Germany (West)	15	21	15	14	11	16
Great Britain	23	21	14	21	14	17
Ireland	16	9	8	13	6	5
Netherlands	1	5	5	2	5	4
Portugal	36	37	29	32	17	24
Switzerland	7	15	5	8	9	16
Africa						
Nigeria	65	22	22	52	30	26

*Respondents were asked: "Here is a list of environmental problems facing many communities. Please tell me how serious you consider each one to be here in your community—very serious, somewhat serious, or not serious at all? a. Poor water quality; b. Poor air quality; c. Contaminated soil; d. Inadequate sewage, sanitation and garbage disposal; e. Too many people, overcrowding; f. Too much noise."

SOURCE: Riley E. Dunlap, George H. Gallup, Jr., and Alec M. Gallup, *Health of the Planet* (Princeton, N.J.: George H. Gallup International Institute, 1993).

TABLE 7.3

PERCEIVED SERIOUSNESS OF ENVIRONMENTAL PROBLEMS IN THE WORLD*
Percentage saying "very serious"

	Air pollution	Water pollution	Contaminated soil	Loss of species	Loss of rain forest	Global warming	Loss of ozone
North America							
Canada	61	77	57	58	71	58	70
United States	60	71	54	49	63	47	56
Latin America							
Brazil	70	69	56	73	77	71	74
Chile	73	77	64	72	71	59	78
Mexico	77	78	77	81	80	62	71
Uruguay	78	77	68	76	80	69	84
East Asia							
Japan	43	43	29	37	47	47	55
Korea (Rep.)	55	49	27	33	24	47	54
Philippines	49	46	42	44	65	40	37
Other Asia							
India	65	50	35	48	54	36	40
Turkey	72	61	54	61	63	45	59
Eastern Europe							
Hungary	54	53	42	47	47	33	47
Poland	77	80	73	76	73	59	66
Russia	71	74	63	61	65	40	59
Scandinavia							
Denmark	61	72	42	62	84	55	65
Finland	58	67	52	48	71	34	60
Norway	69	71	55	61	80	66	70
Other Europe							
Germany (West)	61	70	55	69	80	73	78
Great Britain	52	72	50	60	79	62	66
Ireland	63	74	52	55	67	63	68
Netherlands	30	43	36	45	70	36	47
Portugal	78	81	71	68	82	72	79
Switzerland	62	69	46	61	78	62	68
Africa							
Nigeria	43	44	47	34	31	26	27

*Respondents were asked: "Now let's talk about the world as a whole. Here is a list of environmental issues that may be affecting the world as a whole. As I read each one, please tell me how serious a problem you personally believe it to be in the world—very serious, somewhat serious, not very serious, or not serious at all, or you don't know enough about it to judge? a. Air pollution and smog; b. Pollution of rivers, lakes, and oceans; c. Soil erosion, polluted land, and loss of farmland; d. Loss of animal and plant species; e. Loss of the rain forests and jungles; f. Global warming or the 'greenhouse' effect; g. Loss of ozone in the earth's atmosphere."

SOURCE: Riley E. Dunlap, George H. Gallup, Jr., and Alec M. Gallup, *Health of the Planet* (Princeton, N.J.: George H. Gallup International Institute, 1993).

Among the poll's conclusions:

Environmental problems are rated as one of the three most serious problems in nearly half of the nations surveyed.

Although respondents in the less developed world identify local problems such as air and water quality and lack of sewage treatment facilities as much more serious than residents of the more developed world, both groups rate similarly the seriousness of global problems such as ozone layer depletion, climate change, and loss of biodiversity.

When asked whether the "industrialized or the developing countries are more responsible for today's environmental problems," the most frequently given response in both the wealthy and poor worlds is that both are "equally responsible."

Residents of the less-developed world believe that the most important actions that the more-developed world could take would be to provide educational opportunities, technological assistance, and family-planning services. Support for debt reduction is greater in the LDCs.

There is strong support throughout the world for the creation of an international environmental agency.

The majority of respondents in twenty-one of twenty-four nations indicated that "protecting the environment should be given priority, even at the risk of slowing down economic growth."[26]

Although these results are significant, it is not clear what kinds of specific policy actions are supported by public opinion. Political leadership will play a fundamental role in focusing debate on the choices we face and generating support for the decisions to be made. Tables 7.4, 7.5, and 7.6 provide additional perspective on how willing people are to support changes in environmental policy. The tables demonstrate strong support for accepting responsibility for environmental problems and for the efforts to remedy them.

TABLE 7.4

RESPONSIBILITY FOR THE WORLD'S ENVIRONMENTAL PROBLEMS*
Percentage who think each is responsbile

Economic level (per capita GNP)	Industrial nations	Developing nations	Both equally
Low income			
Nigeria	32	18	37
India	31	12	46
Philippines	30	14	54
Turkey	40	12	39
Poland	45	7	39
Chile	37	10	50
Middle income			
Mexico	37	6	50
Uruguay	38	5	49
Brazil	32	8	56
Hungary	28	9	56
Russia	30	5	57
Portugal	37	3	52
Korea (Rep.)	33	37	23
High income			
Ireland	40	4	46
Great Britain	37	6	50
Netherlands	53	2	40
Canada	37	3	57
United States	29	4	61
Denmark	64	5	27
Germany (West)	54	4	37
Norway	65	3	26
Japan	41	11	28
Finland	58	5	33
Switzerland	46	5	46

*Respondents were asked: "Now, thinking about the world, which do you think is more responsible for today's environmental problems in the world—industrialized countries, developing countries, or do you think they are both equally responsible?"

SOURCE: Riley E. Dunlap, George H. Gallup, Jr., and Alec M. Gallup, *Health of the Planet* (Princeton, N.J.: George H. Gallup International Institute, 1993).

TABLE 7.5

PERCEIVED CONTRIBUTORS TO ENVIRONMENTAL PROBLEMS IN DEVELOPING COUNTRIES*
Percentage who say each contributes "a great deal"

Economic level (per capita GNP)	Consumption by industrialized countries	Multinational companies	Overpopulation in developing countries
Low income			
Nigeria	47	47	55
India	36	30	61
Philippines	44	41	52
Turkey	64	39	52
Poland	25	21	17
Chile	43	37	37
Middle income			
Mexico	55	51	54
Uruguay	48	50	43
Brazil	46	45	37
Hungary	14	13	19
Russia	29	28	18
Portugal	54	51	41
Korea (Rep)	42	41	29
High income			
Ireland	43	43	46
Great Britain	45	43	53
Netherlands	12	50	32
Canada	42	44	50
United States	41	35	50
Denmark	34	35	42
Germany (West)	60	55	62
Norway	57	53	60
Japan	38	25	22
Finland	49	42	57
Switzerland	33	34	52

*Respondents were asked: "How much do you think each of the following contributes to environmental problems in developing countries—a great deal, a fair amount, not very much, or none at all? a. Consumption of the world's resources by industrialized countries; b. Multinational companies operating in developing countries; c. Overpopulation in these developing countries."

SOURCE: Riley E. Dunlap, George H. Gallup, Jr., and Alec M. Gallup, *Health of the Planet* (Princeton, N.J.: George H. Gallup International Institute, 1993).

Public Support for Environmental Protection

TABLE 7.6
SUPPORT FOR CONTRIBUTION TO AN INTERNATIONAL AGENCY AND FOR GIVING AUTHORITY TO THE AGENCY (%)

	Favor contributing money to agency*	Favor giving authority to agency**
Finland	90	74
Netherlands	89	75
Great Britain	89	73
Hungary	84	72
Portugal	83	74
Korea (Rep.)	83	74
Norway	83	65
Germany(West)	82	78
Russia	79	76
Ireland	79	7
Switzerland	79	71
Japan	78	65
Denmark	78	52
Poland	78	7
Mexico	77	61
Canada	77	70
India	75	57
Philippines	75	64
Turkey	75	60
United States	74	63
Nigeria	73	70
Chile	68	54
Uruguay	61	54
Brazil	56	62

*Respondents were asked: "Would you favor or oppose our government contributing money to an international agency to work on solving global environmental problems—strongly favor, somewhat favor, somewhat oppose, or strongly oppose?"

**Respondents were asked: "Would you favor or oppose giving an international agency the authority to influence our government's policy in environmentally important areas—strongly favor, somewhat favor, somewhat oppose, or strongly oppose?"

SOURCE: Riley E. Dunlap, George H. Gallup, Jr., and Alec M. Gallup, *Health of the Planet* (Princeton, N.J.: George H. Gallup International Institute, 1993).

A New Environmental Paradigm

A political system that is incremental in nature, is largely an attempt to muddle through problems. It fosters stability and reflects the preferences of those who have political power. It is likely to avoid hasty changes or misconceived dramatic experiments in policy innovation. But incrementalism does not encourage examination of basic policy direction or changes to it when necessary. Fundamental problems and basic conflicts among values are not likely to be addressed. The future is discounted; short-run calculations dominate.[27]

Economists have traditionally ignored the costs of pollution and depletion of scarce resources. They usually argue that markets will respond to scarcity by raising prices and encouraging conservation or developing substitutes. But such essential resources as air have no market price. Herman Daly believes that if we more carefully calculate the true costs of economic activity, we will find that economic growth may come at a very high price. It is possible, Daly warns, that "economic growth increases environmental costs faster than it increases production benefits and thereby makes us poorer rather than richer." Daly argues for several policy changes: economic statistics that account for depletion of natural resources should be developed; nondepletable resources should be consumed only at the same rate alternatives are developed; and taxes should be shifted toward resource consumption and pollution production.[28]

In 1991, Lawrence Summers, then the World Bank's chief economist, wrote a memo arguing that the bank should encourage migration of dirty industries to the developing world. Since pollution is measured by the cost of wages lost due to illness and premature death, pollution in low-wage countries makes economic sense. Since pollution control costs are high in countries where pollution levels are high, industry should be encouraged to move to underpolluted countries of the South. Since people in poor countries have relatively short life spans, they should not be worried about increased exposure to toxic chemicals. Summers claimed that his comments were meant to be "highly ironic," but they provide a stark example of the nature of economic reasoning.

A New Environmental Paradigm

They also reflect the kind of thinking that seems to lie behind World Bank lending practices. How else to explain projects that develop fossil fuels, construct large dams that displace local residents, direct only a tiny fraction of funds to energy conservation, and finance roads through ecologically sensitive lands?[29]

Deep ecology advocates focus on the preservation of biodiversity as the primary value: Human wants and interests must give way to the survival of all species. Human society must be restructured to harmonize with natural conditions and constraints.[30] Social theorists claim that ecological and social problems are inextricably intertwined. Discrimination, domination, and abuse are social as well as environmental phenomena; the solution to both kinds of problems lies in a radical economic, social, and political restructuring aimed at justice, freedom, and equality.[31]

Norman Myers argues that environmental risks and the possibility of environmental collapse should be understood as one of the greatest threats to our collective security we face. Deforestation and other environmentally unsustainable actions by poor farmers in poor countries in their struggle to survive make their long-term prospects even bleaker. Lack of assistance in remedying their problems attracts them to rural guerrilla and resistance movements that result in conflict. Environmental decline causes agricultural productivity to fall; food shortages trigger social unrest, disorder, and war. Although these problems primarily concern the less-developed world, the wealthy world also faces conflict over declining fisheries, water supplies, and other natural resources. Myers believes environmental conflicts and competition over natural resources will dominate the next forty years, just as the cold war dominated the past forty years. But the end of the cold war also creates the opportunity for collective efforts to pursue our security. The less- and more-developed worlds are inextricably linked; and environmental collapse in one area involves the rest of the world.[32]

Matthew Cahn argues that liberal society's emphasis on self-interest, the protection of individual rights, and capitalism's constant drive for expansion, growth, and consumption can hamper environmental preservation efforts. Policy makers have understated the challenges in protecting environmental quality as they fail to address head on the tension between liberalism and environmentalism. Cahn contends that liberalism prevents us from effectively protecting environmental quality, and that to achieve this goal we will need to replace our narrow model of self-interest with a broader approach such as that suggested by communitarians.[33]

Immanuel Wallerstein argues that liberalism has been a compromise view between conservatism and socialism. It has been an attractive option because it promises unlimited growth, individual freedom, unrestrained consumption. But those assumptions/values are no longer tenable in the face of environmental pollution, risks, limits, and other social stresses. We need an alternative vision that fosters decentralized, democratic organizations aimed at liberating human potential rather than simply pursuing economic growth.[34]

What are the prospects for developing a new paradigm that is rooted in ecology?[35] William Ophuls and A. Stephen Boyan argue that working with the current paradigm to prevent further decline is essential, but not sufficient: "To accept current political reality as not itself subject to radical change is to give away the game at the outset and render the situation hopeless by definition." Politics is not just the "art of the possible," but also the *"art of creating new possibilities* for human progress." Solutions will necessarily be branded as unrealistic, since they will naturally conflict with current political and economic values and expectations. Until real change is possible, environmental policy will be little more than an effort to keep conditions from worsening.[36]

The transition to a new paradigm will take several decades. We do not know now exactly what political changes will need to be made, and policy will have to evolve as our experience and knowledge increase.[37] But we need to make some changes now, and move, albeit incrementally, in new directions. We also need to commit now to find out what else we need to do. However, it is "probably much too late for a carefully planned transition to a steady state," Ophuls and Boyle conclude.

> We are so committed to most of the things that cause or support ecological evils that we are almost paralyzed; nearly all the con-

structive actions that could be taken at present (for example, drastically restricting population growth) are so painful to so many people in so many ways that they are indeed totally unrealistic, and neither politicians nor citizens would tolerate them. Only after nature has mandated certain changes and overwhelmingly demonstrated the advisability of others will it be possible to think in terms of a concrete program of transition.[38]

One element of a new paradigm might be to broaden our notion of community to include the ecosystems around us. Herman E. Daly and John B. Cobb argue that fundamentally, people are not autonomous individuals but persons in a broader community, which shapes their identity and gives meaning to their lives. A "biocentric" vision must be central to our political economy; it must go beyond an anthropocentric view of the environment to a deep caring for the biosphere. Social decisions must be made "with the health of the biosphere in view."[39] Aldo Leopold, in *A Sand County Almanac*, comes to recognize that

> the individual is a member of a community of interdependent parts. His instincts prompt him to compete for his place in the community, but his ethics prompt him also to cooperate . . . . The land ethic simply enlarges the boundaries of the community to include soils, waters, plants, and animals, or collectively: the land.[40]

One way of thinking about the interconnectedness of the global environment is the idea of Gaia.[41] Gaia is an ancient Greek goddess, the "earthmother who brought forth the world and the human race from the 'gaping void, Chaos.'" Feminists reclaimed her as a symbol for a new earth-based approach to spirituality and ecology. Gaia became the female member of the godhead, giving women equal status with men and a powerful symbol of life.[42] In 1977, the British scientist James Lovelock borrowed the term in proposing a theory of the earth as a living organism. His book, *Gaia: A New Look at Life on Earth*, argued that all living matter on earth was part of a single, living, self-regulating entity whose characteristics are greater than simply the sum of its constituent parts.[43] Gaia meant that the earth, as a living organism, could and would make whatever changes and adjustments in its atmosphere necessary to survive. Feminist spirituality and biological theory became powerfully linked in providing a new image to guide thinking about the environment. Scientific conferences, environmental meetings, music, art, and pop culture all extolled Gaia's virtues as a new way of thinking about and understanding the earth.

The idea of Gaia, however, seems to raise as many questions as it answers. Scientists have argued that if Gaia is a self-regulating system, it may correct human-caused problems by gradually eliminating their source. Feminists have asked whether Gaia reinforces a deeply held view of men in Western society that "Mom always comes along and cleans up after them."[44] For some, ecofeminism is a more attractive idea that makes strong connections between women and nature and argues that liberating women and nature are central to environmental preservation. Carolyn Merchant writes that one version, liberal ecofeminism, seeks to "alter human relations with nature from within existing structures of governance through the passage of new laws and regulations." Another version, cultural ecofeminism, "analyzes environmental problems from within its critique of patriarchy and offers alternatives that could liberate both women and nature." Socialist ecofeminism is rooted in an assessment of capitalist patriarchy, focusing on how the "patriarchal relations of reproduction reveal the domination of women." From such perspectives, women see pollution not as the byproduct of industrial society but as "assaults on their own bodies and on those of their children."[45]

Some ecofeminists champion an environmental ethic of care and nurture that comes from women's "culturally constructed experiences." Environmental preservation would be guided not by rights and rules but by pluralistic approaches that recognize the multiple voices of women and are rooted in love, trust, and care. Merchant proposes a partnership ethic that treats humans, including male and female partners, "as equals in personal, household, and political relations" and humans as equal partners with "nonhuman nature."

A New Environmental Paradigm

Just as human partners "must give each other space, time, and care" in order to flourish, humans "must give nonhuman nature space, time, and care, allowing it to reproduce, evolve, and respond to human actions." This requires avoiding actions, such as cutting down trees and damming rivers, that result in displacing wildlife and making humans more susceptible to flooding; using restraint in introducing new technologies and chemicals; and allowing room in planning human development for ecological surprises.[46]

Vice President Al Gore has become perhaps the most visible proponent of a new political economy based on moral commitments. The global environmental crisis is a manifestation of a deeper spiritual crisis—a failure to take responsibility for our actions. For Gore, incremental changes are not enough: "We must make the rescue of the environment the central organizing principle for civilization."[47] His agenda for change includes a new kind of "*eco*-nomics," one that involves reorienting economic activity toward environmentally sensitive production and investment in future resources, rather than relying on such technological fixes as genetic engineering or desalinization of water. He proposes a global Marshall Plan to stabilize population growth, develop and use "environmentally appropriate technologies," negotiate new international treaties, and increase education about global environmental issues. A "Strategic Environmental Initiative" would provide for:

1. Tax incentives for research and development to encourage new developments;

2. Government purchases of cleaner technologies and recycled materials to stimulate such developments;

3. Guaranteed profits for new, environment-friendly technologies;

4. New assessments of the economic and ecological impact of proposed technologies;

5. Training centers for educating an elite group of environmental planners and technicians;

6. Export controls on technologies and products that are environmentally harmful;
7. Greater legal protection for investments.[48]

The policy agenda is comprehensive and pervasive, but most important, it would mark a fundamental change in our moral commitments to each other and to those who come after us.

Vaclav Smil argues that social, economic, political, and intellectual flexibility is critical as we adapt to changes in the global environment. Neither pessimism nor optimism, neither confidence nor despair, are warranted as we come to understand the tremendous environmental threats confronting us. Instead, we need hope, a sense of possibility. Scientific knowledge is an uncertain guide, even in the most mundane tasks, such as determining how much fertilizer to apply. Modeling of environmental conditions is extremely limited; we lack the computing power to make models specific enough to be very accurate. We often make meaningless generalizations that tell us little about the real nature of problems or how to solve them. We fail to recognize the unpredictability of human events: Randomness and indeterminacy are more realistic than are simplistic causal explanations.[49]

For Smil, these changes require a great deal of new thinking—thinking that will transform the management of organizations as well as individuals' goals and choices. We will not find solutions to global problems in dramatic new and highly sophisticated technologies, but in immediate measures that restrain consumption and permit people in the less developed world to raise their quality of life without further damaging the environment. This will require a new economics and a new commitment to social governance, centering on better market approaches, more responsive local and national institutions. A strong commitment to international cooperation will eliminate excessive consumption, use the most efficient regulatory techniques, and pursue consensual management.

Many writers have concluded that religious and spiritual values are essential in constructing a green public philosophy. Robert Booth Fowler, for example, describes the way in which Protestant churches, elites, and members have embraced environmental issues and ideas during the past twenty-five years. He argues that Protestantism encompasses a great diversity of thought concerning the environment, ambivalence about the role of politics in responding to environmental challenges, and a strong commitment to the idea of community and a holistic natural world. Fowler's analysis also goes beyond environmentalism to explore broader issues of community and the impact of Protestant political thought on that broader debate over public life and the public good.[50]

Religious traditions emphasize respect for all forms of life as a manifestation of respect for the creator. Caring for all creation means protecting endangered species as well as indigenous peoples who die as their homelands are destroyed.[51] Religion has played a powerful role in providing ethical guidance in many areas of life. It can do so in this area as it fosters an "environmentalism of the spirit" that requires that we care for the earth as God's creation.[52]

Ophuls and Boyan's prescription is ultimately rooted in a similar concern. The steady state will "almost certainly have a religious basis—whether it be Aristotelian political and civic excellence, Christian virtue, Confucian rectitude, Buddhist compassion, Amerindian love for the land, or an amalgam of these and other spiritual values."[53] They maintain that the ecological crisis

> is fundamentally a moral and spiritual crisis. . . . The virtues that have been championed in philosophy and religion for ethical and spiritual reasons will become the central means of survival . . . . Political and spiritual wisdom alike urge the adoption of the minimal, frugal, steady state as the form of a post-industrial society.[54]

Wendell Berry has explored the possibilities of a Christian ethic of caring for the earth. He recognizes the "culpability of Christianity in the destruction of the natural world."[55] Most Christian organizations are indifferent to "the ecological, cultural, and religious implications of industrial economics"; missionaries and mercenaries who seek for wealth seem to have the same motivation.[56] Nevertheless, Berry finds in biblical teachings the elements of a new paradigm. Humans do not own the earth or any part of it: "The earth is the Lord's, and the fulness thereof: the world and they that dwell therein" (Psalms 24:1). "All things were made by [God]; and without him was not anything made that was made" (John 1:3). For Berry, the "creation is not in any sense independent of the Creator, the result of a primal creative act long over and done with, but is the continuous, constant participation of all creatures in the being of God."[57] The destruction of nature is "blasphemy," since we "have the right to use the gifts of nature but not to ruin or waste them. We have the right to use what we need but no more."[58] Christianity has in it the seeds of a new global paradigm. Indeed, as Berry argues, its founder challenged the established powers and ways of thinking and acting. But most of Christianity has stood by while the pursuit of wealth has ravaged the world, and assumed with others that such "progress" is good.[59]

The concerns of indigenous peoples are a central part of global environmental efforts. The first revised text of the draft *Universal Declaration on the Rights of Indigenous Peoples*, calls on nations to "comply with and effectively implement all international human rights instruments as they apply to indigenous peoples," including their land rights:

> The right of collective and individual ownership, possession and use of the lands or resources which they have traditionally occupied or used. The lands may only be taken away from them with their free and informed consent as witnessed by a treaty or agreement....
>
> The right to require that States consult with indigenous peoples and with both domestic and transnational corporations prior to the commencement of any large-scale projects, particularly natural resource projects or exploitation of mineral and other subsoil resources, in order to enhance the projects' benefits and to mitigate any adverse economic, social, environmental and cultural effect. Just and fair compensation shall be provided for any such activity or adverse consequence undertaken.[60]

Delegates to the United Nations Conference on Environment and Development adopted as part of Agenda 21 a list of commit-

A New Environmental Paradigm

ments participating nations agreed to make concerning the rights of indigenous peoples.[61] Among these commitments were the following provisions designed to empower native peoples:

> Recognition that the lands of indigenous people and their communities should be protected from activities that are environmentally unsound or that the indigenous people concerned consider to be socially and culturally inappropriate;
>
> Recognition of their values, traditional knowledge and resource management practices with a view to promoting environmentally sound and sustainable development;
>
> Recognition that traditional and direct dependence on renewable resources and ecosystems, including sustainable harvesting, continues to be essential to the cultural, economic and physical well-being of indigenous people and their communities;
>
> Enhancement of capacity building for indigenous communities, based on the adaptation and exchange of traditional experience, knowledge and resource-management practices, to ensure their sustainable development; and
>
> Establishment, where appropriate, of arrangements to strengthen the active participation of indigenous people and their communities in the national formulation of policies, laws and programmes relating to resource management and other development processes that may affect them, and their initiation of proposals for such policies and programmes.[62]

Governments should, in "full partnership with indigenous people"

> develop or strengthen national arrangements to consult with indigenous people and their communities with a view to reflecting their needs and incorporating their values and traditional and other knowledge and practices in national policies and programmes in the field of natural resource management and conservation and other development programmes affecting them.[63]

Since the International Court of Justice relies on principles of equity in deciding cases, Edith Brown Weiss has proposed that these "equitable standards for the allocation and sharing of resources and benefits" can be used in developing specific principles of intergenerational equity. Among such principles are the following:

> Each generation should be required to conserve the diversity of the natural and cultural resource base so that it does not unduly restrict the options available to future generations in solving their problems and satisfying their own values.
>
> Each generation should be required to maintain the quality of the planet so that it is passed on in no worse condition than that in which it was received.
>
> Each generation would provide its members with equitable rights of access to the legacy of past generations and should conserve this access for future generations.[64]

These principles do not require that the natural environment remain unchanged; they can be ensured through technological innovation and balancing the needs of current and future generations. They emphasize equality among generations and grant to future generations flexibility in achieving their own goals through their own values. To be successful, they must be generally shared and widely acceptable.[65]

Intergenerational planetary rights are group rights, recognized for all members of each generation. All decisions we make can be judged from the basis of their impact on future generations. International law can provide a legal underpinning of sustainable development. An alternative view points us backward, where the members of the current generation owe a debt to previous generations, which they pay to succeeding generations in the form of investments. The obligations of the current generation are a result of the benefits they have received from their ancestors, rather than a normative view concerning stewardship and equality among generations.[66] Strategies for implementing intergenerational equity include representation of future generations in decision making by specifying agents to represent those interests, employing natural resource accounting, requiring assessments of the impact of proposed actions on future generations, and articulating and emphasizing rights and norms of intergenerational equity. This requires a new global ethic of responsibility and equity that must begin with a major educational effort. Our institutions are oriented toward short-term considerations; how can we reorient them toward a long-term perspective?[67]

The theory of intergenerational equity holds that humans share the planet "in common with all members of our species: past generations, the present generation, and future generations."[68] The key relationships are between the current generation and nature, and between different generations: "If one generation degrades the environment severely, however, it will have violated its intergenerational obligations relating to the care of the natural system."[69] The human community is a "partnership among all generations."[70] Since we do not know how many generations there will be in the future, or how many members in each generation, we can take the position similar to the original position first proposed by John Rawls, of a "generation that is placed somewhere along the spectrum of time, but does not know in advance where it will be located. Such a generation would want to inherit the earth in at least as good condition as it had been for any previous generation and to have as good access to it as previous generations have had."[71] This view of intergenerational equity is rooted in the teachings of the world's great religions. It is reflected in the key documents of international law that champion the rights of all humans in all eras—the United Nations Charter, the preamble to the Universal Declaration of Human Rights, and other documents. It is expressly addressed in the preamble to the Stockholm Declaration on the Human Environment, where the purpose of the conference was stated as defending and protecting the environment for "present and future generations."[72]

The arguments aimed at protecting the rights and interests of future generations can also be applied to those now living in the less-developed world. Proposals for reconceptualizing the global political economy and making environmental preservation the primary value all promise to remedy global inequality as well. Demands for representation of the interests of future generations, in particular, can and should be integrated with the compelling call for efforts to remedy the injustices and suffering that plague so many of the current inhabitants of the global community. We cannot remedy global environmental threats without a strong commitment to addressing the problems that face the most vulnerable of our fellow human beings.[73] "Sustainable development does not just mean conserving present resources for future generations," argued one spokesperson for the Third World Network (a group of some two hundred NGOs) at the 1992 United Nations Conference on the Environment and Development;. It also means "reducing the excessive consumption of a minority so that resources are 'freed' to meet the basic needs of the rest of the humanity within this generation."[74]

As we develop that commitment, we can begin to look beyond the modest policies discussed earlier to more fundamental changes in international trade, the behavior of multinational corporations, consumption, and production. A theory of inter- and intragenerational equity can set a standard for formulating and implementing global agreements. Now, though, it bumps up against the realities of economic and political power. It will not likely have a great impact on nations unless it becomes a central part of the agenda of institutions responsible for the formulation and implementation of global accords.

Theories of international law can provide a banner around which appeals to enlightened self-interest can be made in formulating and implementing global agreements. International judicial bodies can make them effective in specific cases.

The metaphor of a road is often used to represent policy choices. A choice has important consequences: it may mean a path to ecologically sustainable activity, or one that increases environmental degradation. Ed Ayres, editor of *World Watch* magazine, argues that the metaphor is problematic. A road is "linear, rigid . . . a highly artificial imposition on the planet" that implies an "incremental, single-minded, conquer-the-earth progress." We need a new

A New Environmental Paradigm

metaphor for public choices, such as a leaping frog, that encourages us to think of ways to "skip the mistakes of the past" and "change quickly rather than ponderously." We need to bypass old ways of doing things: instead of making slow turns, we need to leap ahead to employ new technologies and new ways of acting.[75]

Can we expect attitudes and behaviors to change as people come to understand the prerequisites of a sustainable environment? People do not necessarily change their behavior even when they are aware of the seriousness of a potential threat. Many people have little information about the nature of the global environmental crisis:

> How reasonable is it to expect from the public at large a sophisticated ecological understanding any time soon, especially when the academic, business, professional, and political elites who constitute the so-called attentive and informed public show little sign of having understood, much less embraced, the ecological world view?[76]

Perhaps we can come to rely more on technological fixes to solve our problems. Even if this is true, however, it poses a profound political challenge. The more sophisticated the technology, the greater the role for those with expertise, and the more difficult it will be to engage in democratic deliberation and accountability for decision making. Even with increased reliance on experts, the choices will continue to be vexing, political ones concerning the costs and benefits, the distribution of risks, and the insurance against uncertainty. The decisions confronting us, such as whether to expand our reliance on nuclear power, are fraught with uncertainties. Technical expertise alone will not provide the kind of moral leadership needed to deal with them.[77] There is no substitute for public debate and education. As we develop our capacity for discourse and our commitment to discussion, we will strengthen our capacity to devise the solutions to environmental problems and to foster the kind of support for collective action that will be required.

References

1. Government of Canada, Canada's National Report, United Nations Conference on Environment and Development (August 1991), 84.
2. Alan Thein Durning, "Supporting Indigenous Peoples" in Lester Brown et al., State of the World 1993 (Washington, D.C.: Worldwatch Institute, 1993), 92.
3. Al Gore, *Earth in the Balance: Ecology and the Human Spirit* (New York: Houghton Mifflin, 1992), 240.
4. For a review of the debate over the ecological and health and risks of pollution, see Conservation Foundation, *State of the Environment: A View toward the Nineties* (Washington, D.C.: Conservation Foundation, 1987); Lester Lave, *The Strategy of Social Regulation* (Washington, D.C.: Brookings Institution, 1981); National Research Council, *Risk Assessment in the Federal Government: Managing the Process* (Washington, D.C.: National Academy Press, 1983); Edith Effron, *The Apocalyptics: Cancer and the Big Lie* (New York: Simon & Schuster, 1984); and H. W. Lewis, *Technological Risk* (New York: Norton, 1990).
5. National Research Council, *Risk Assessment in the Federal Government: Managing the Process* (Washington, D.C.: National Academy Press, 1983), 21.
6. For a discussion of these issues, see, generally, Michael Kraft and Norman Vig, *Technology and Politics* (Durham, N.C.: Duke University Press, 1988).
7. For an elaboration of these and related themes, see David Dickson, *The New Science of Politics* (Chicago: University of Chicago Press, 1989).
8. For a thoughtful discussion of the hazards of risk communication, see Conservation Foundation, *Risk Communication* (Washington, D.C.: Conservation Foundation, 1987).
9. The debate over policy options to address acid rain in the U.S. provides examples of both approaches. Some Reagan administration officials clearly did not want to impose new regulatory burdens, hence the call for more research as a delaying tactic. Others believed that policies should not be fully designed until the federal government completed its decade-long assessment of acid rain, slated to conclude in 1990. For more on that view, see Michael S. McMahon, "Balancing the Interests: An Essay on the Canadian-American Acid Rain Debate," in John E. Carroll, ed., *International Environmental Diplomacy* (New York: Cambridge University Press, 1988), 147–71.
10. Aaron Wildavsky, "The Secret of Safety Lies in Danger," in Gary Bryner and Dennis Thompson, eds., *The Constitution and the Regulation of Society* (Ithaca, N.Y.: SUNY Press/Brigham Young University Press, 1988), 51–52.

11. David Braybrooke and Charles E. Lindblom, *A Strategy of Decision* (New York: Free Press, 1963); Joseph G. Morone and Edward J. Woodhouse, *Averting Catastrophe: Strategies for Regulating Risky Technologies* (Berkeley: University of California Press, 1986); Joseph G. Morone and Edward J. Woodhouse, *The Demise of Nuclear Energy: Lessons for Democratic Control of Technology* (New Haven: Yale University Press, 1989).
12. Stephen H. Schneider, *Global Warming: Are We Entering the Greenhouse Century?* (San Francisco: Sierra Club Books, 1989), 283–85.
13. Daniel Bodansky, "Scientific Uncertainty and the Precautionary Principle," *Environment* 33 (September 1994); Robert Costanza and Laura Cornwell, "The 4P Approach to Dealing with Scientific Uncertainty," *Environment* 34 (November 1992): 12–20, 42.
14. U.S. Council on Environmental Quality, *Environmental Quality: 15th Annual Report of the Council on Environmental Quality* (Washington, D.C.: U.S. GPO, 1984), 226–27.
15. Frederick Allen, "The Situation: What the Public Believes; How the Experts See It," *EPA Journal* (November 1987): 9–12.
16. For further discussion of risk assessment in the U.S., see Elizabeth L. Anderson, and the Carcinogen Assessment Group of the U.S. Environmental Protection Agency, "Quantitative Approaches in Use to Assess Cancer Risk," *Risk Analysis* 3, no. 4 (1983): 277–295; Robert W. Crandall and Lester B. Lave, eds., *The Scientific Basis of Health and Safety Regulation* (Washington, D.C.: The Brookings Institution, 1981); David Doniger, "The Gospel of Risk Management: Should We Be Converted?" *Environmental Law Reporter* 14 (June 1984): 10222; Howard Latin, "Good Science, Bad Regulation, and Toxic Risk Assessment," *Yale Journal on Regulation* 5 (1988): 89; John M. Mendeloff, *The Dilemma of Toxic Substance Regulation, How Overregulation Causes Underregulation at OSHA* (London: MIT Press, 1988); Albert L. Nichols, and Richard J. Zeckhauser, "The Perils of Prudence: How Conservative Risk Assessments Distort Regulation," *Regulation* (November/December 1986): 13–24; William D. Ruckelshaus, "Risk in a Free Society," *Environmental Law Reporter* 14 (June 1984): 10190; Milton Russell and Michael Gruber, "Risk Assessment in Environmental Policy Making," *Science* 236 (17 April 1987): 286–290; Leonard A. Sagan, "Beyond Risk Assessment," *Risk Analysis* 7, no. 1 (1987): 1–2; U.S. Environmental Protection Agency, *Guidelines for Performing Regulatory Impact Analysis*, EPA-230-01-84-003 (December 1983).
17. Peter Sandman, "Risk Communication: Facing Public Outrage" *EPA Journal* (November 1987): 21–22.
18. Barry Lopez, *The Rediscovery of North America* (Lexington, Ky.: University Press of Kentucky, 1990), 16.

19. See, generally, United States Catholic Conference, "Renewing the Earth," 14 November 1991, 2.
20. Gregg Easterbrook, *A Moment on the Earth: The Coming Age of Environmental Optimism* (New York: Viking Penguin, 1995), xvi.
21. Julian Simon, "Pre-Debate Statement," in Norman Myers and Julian L. Simon, *Scarcity or Abundance? A Debate on the Environment* (New York: Norton, 1994), 5–22.
22. Tom Tietenberg, *Environmental and Natural Resource Economics*, 3d ed. (New York: Harper Collins, 1992), 356–57.
23. Thomas F. Homer-Dixon, quoted in William K. Stevens, "Feeding a Booming Population Without Destroying the Planet," *New York Times*, 5 April 1994.
24. Anne E. Platt, "It's About More Than Sea Cucumbers," *World Watch* (May/June 1995): 2.
25. Riley E. Dunlap, George H. Gallup Jr., and Alec M. Gallup, *Health of the Planet* (George H. Gallup International Institute, Princeton, NJ, 1993).
26. Ibid.
27. William Ophuls and A. Stephen Boyan, Jr., *Ecology and the Politics of Scarcity Revisited* (New York: W. H. Freeman, 1992), 245–46.
28. Quoted in G. Pascal Zachary, "A 'Green Economist' Warns Growth May be Overrated," *The Wall Street Journal*, 25 June 1996, B1.
29. See Paul Rauber, "World Bankruptcy," *Sierra* (July/August 1992): 35–38.
30. See Dave Foreman, *Confessions of an Eco-Warrior* (New York: Crown, 1991).
31. See Murray Bookchin, *Remaking Society: Pathways to a Free Future* (Boston: South End Press, 1990).
32. Norman Myers, *Ultimate Security: The Environmental Basis of Political Stability* (New York: Norton, 1995).
33. Matthew Alan Cahn, *Environmental Deceptions: The Tension Between Liberalism and Environmental Policymaking in the United States* (Albany: State University of New York Press, 1995).
34. Immanuel Wallerstein, *After Liberalism* (New York: New Press, 1996).
35. For more on alternative paradigms, see Mary E. Clark, *Ariadne's Thread: The Search for New Modes of Thinking* (New York: St. Martins Press, 1989); Al Gore, *Earth in the Balance: Ecology and the Human Spirit* (Boston: Houghton Mifflin, 1992); David Ray Griffin and Richard Falk, eds., *Postmodern Politics for a Planet in Crisis* (Albany: State University of New York Press, 1993); Robert L. Heilbroner, *An Inquiry into the Human Prospect* (New York: Norton, 1991); Alexander King and Bertrand Schneider, *The First Global Revolution* (New York: Pantheon, 1991); Chris Maser, *Global Imperative: Harmonizing Culture*

and Nature (Walpole, N.H.: Stillpoint Publishing, 1992); Carolyn Merchant, *Radical Ecology: The Search for a Liveable World* (New York: Routledge, 1992); Lester W. Milbrath, *Envisioning a Sustainable Society: Learning Our Way Out* (Albany: State University of New York Press, 1989); William Ophuls and A. Stephen Boyan, Jr., *Ecology and the Politics of Scarcity* (New York: W. H. Freeman, 1992); Norman Myers, *Ultimate Security: The Environmental Basis of Political Stability* (New York: W.W. Norton, 1993); David W. Orr, *Ecological Literacy: Education and the Transition to a Postmodern World* (Albany: State University of New York, 1992); Mark Sagoff, *The Economy of the Earth: Philosophy, Law, and the Environment* (Cambridge: Cambridge University Press, 1988); Allan Schnaiberg and Kenneth Alan Gould, *Environment and Society: The Enduring Conflict* (New York: St. Martin's, 1994); Christopher D. Stone, *The Gnat Is Older than Man: Global Environment and Human Agenda* (Princeton, N.J.: Princeton University Press, 1993); and Jonathan Weiner, *The Next Hundred Years: Shaping the Fate Of Our Living Earth* (New York: Bantam, 1990).

36. Ophuls and Boyan, *Ecology and the Politics of Scarcity Revisited*, 282.
37. For discussions of global environmental politics see Ann Bramwell, *The Fading of the Greens: The Decline of Environmental Politics in the West* (New Haven, Conn.: Yale University Press, 1994): Tony Brenton, *The Greening of Machiavelli: The Evolution of International Environmental Politics* (London: Earthscan, 1994); Lynton Keith Caldwell, *International Environmental Policy: Emergence and Dimension*, 2d. ed. (Durham, NC: Duke University Press, 1990); James E. Harf and B. Thomas Trout, *The Politics of Global Resources: Energy, Environment, Population, and Food* (Durham, NC: Duke University Press, 1986) Andrew Hurrell and Benedict Kingsbury, *The International Politics of the Environment* (Oxford: Clarendon Press, 1992); Sheldon Kamieniecki, *Environmental Politics in the International Arena* (Albany, N.Y.: State University of New York Press, 1993); Ronnie D. Lipschutz and Ken Conca, eds., *The State and Social Power in Global Environmental Politics* (New York: Columbia University Press, 1993); John McCormick, *Reclaiming Paradise: The Global Environmental Movement* (Bloomington: Indiana University Press, 1989); Jonathon Porritt, *Seeing Green: The Politics of Ecology Explained* (Oxford: Basil Blackwell, 1985); Gareth Porter and Janet Welsh Brown, *Global Environmental Politics* (Boulder, Colo.: Westview Press, 1991); Charlene Spretnak and Fritjof Capra, *Green Politics: The Global Promise* (Santa Fe, N.M.: Bear and Company, 1986); Lawrence E. Susskind, *Environmental Diplomacy: Negotiating More Effective Global Agreements* (New York: Oxford University Press, 1994); Ernst Ulrich von Weizacker, *Earth Politics* (London: Zed Books, 1994); and John Young, *Sustaining the Earth: The Story of the Environ-*

mental Movement—Its Past Efforts and Future Challenges (Cambridge, Mass.: Harvard University Press, 1990).
38. Ophuls and Boyan, *Ecology and the Politics of Scarcity Revisited*, 283.
39. Herman E. Daly and John B. Cobb, *For the Common Good: Redirecting the Economy Toward Community, the Environment, and a Sustainable Future* (Boston: Beacon Press, 1989), 202.
40. Aldo Leopold, *A Sand County Almanac* (New York: Ballantine Books, 1970), 239.
41. These three paragraphs are taken from Jacqueline Switzer with Gary Bryner, *Environmental Politics* (New York: St. Martin's Press, 1997).
42. See Charlene Spetnak, *Lost Goddesses of Early Greece: A Collection of Pre-Hellenic Mythology* (Ann Arbor, Mich.: Moon Books, 1992), 30–31; quoted in Carolyn Merchant, *Earthcare: Women and the Environment* (New York: Routledge, 1995), 3.
43. James Lovelock, *Gaia: A New Look at Life on Earth* (New York: Oxford University Press, 1979).
44. Quoted in Merchant, *Earthcare*, 5.
45. Ibid., 5–6.
46. Ibid., 7–8.
47. Gore, *Earth in the Balance*, 269.
48. Ibid., 320.
49. Vaclav Smil, *Global Ecology: Environmental Change and Social Flexibility* (London: Routledge, 1993).
50. Robert Booth Fowler, *The Greening of Protestant Thought* (Chapel Hill: University of North Carolina Press, 1995).
51. Sandman, "Risk Communication," 7–9.
52. Gore, *Earth in the Balance*, 238–65.
53. Ophuls and Boyan, *Ecology and the Politics of Scarcity Revisited*, 292.
54. Ibid., 299, 302.
55. Wendell Berry, *Sex, Economy, Freedom and Community* (New York: Pantheon, 1993), 93.
56. Ibid., 94.
57. Ibid., 97.
58. Ibid., 98.
59. Ibid., 115.
60. First revised text of the draft *Universal Declaration on the Rights of Indigenous Peoples*, quoted in Robert N. Clinton, Nell Jessup Newton, and Monroe E. Price, *American Indian Law* (Charlottesville, Va.: Michie, 1991), 166–71.
61. United Nations Conference on Environment and Development, *Agenda 21*, Chapter 26, "Recognising and Strengthening the Role of Indigenous People and Their Communities."
62. Ibid., Chapter 26, at 26.4.

References

63. Ibid., Chapter 26, at 26.7.
64. Edith Brown Weiss, "Intergenerational Equity: Toward an International Legal Framework," in Nazli Choucri, ed., *Global Accord: Environmental Challenges and International Responses* (Cambridge, Mass.: MIT Press, 1993), 333–53, at 342.
65. Ibid., at 343.
66. See Jerome Rothenberg, "Economic Perspectives on Time Comparisons: Alternative Approaches to Time Comparisons," in Nazli Choucri, ed., *Global Accord*, 355–397.
67. Weiss, "Intergenerational Equity," at 348–49, 353.
68. Ibid., at 334.
69. Ibid., at 335.
70. Ibid., at 335.
71. Ibid., at 335.
72. Ibid., at 337.
73. See Tom Athanasiou, *Divided Planet: The Ecology of Rich and Poor* (Boston: Little, Brown, 1996).
74. Quoted in Mark Hertsgaard, "The View from 'El Centro del Mundo,'" *Amicus Journal* (Fall 1992): 13.
75. Ed Ayers, "Note to Readers," *World Watch* (September/October 1996): 3.
76. Ophuls and Boyan, *Ecology and the Politics of Scarcity Revisited*, 203.
77. Ibid., 211, 215.

Index

NOTE: Italicized page numbers refer to boxes and tables.

acid rain, 4, 6, 24, 76*n*21, 338*n*9
   and Clean Air Act, 143–44
   and emissions trading, 143–44, 155–56
   and integration of environmental and other policies, 107
   international agreements about, 24–29, *25*, 73, 107, 126
   and international law, 126
   and market-based incentives, 143–44
   See also specific agreement
adaptation, 133–34, 257–58, 259, 311
adjustment, promoting, 106, 123–27
Administrative Committee on Coordination (UN), *181*, *183*, 188
Advisory Committee for the Coordination of Information Systems (ACCIS) (UN), *185*, 207
Advisory Committee on Science and Technology for Development (ACAST) (UN), *185*
Advisory Committee for Trade Policy and Negotiations (U.S.), 226
Africa, *21*, *52*, 71–72, *178*, 203.
   See also specific agreement
African Development Bank, *178*, 203

Agency for International Development (AID), 277
Agenda 21 (1992), 16–18, 19, *22*, 188, 202, 203, 283, 284, 332–33
agenda
   of "Earth Summit," 16–19, 87, 88
   global, 263–64
Agreement for Cooperation in Dealing with Pollution of the North Sea Oil (1969), *20*
Agreement Regarding Monitoring of the Stratosphere (Paris, 1976), *30*
Agreement to Ban Ozone-Depleting Substances (London, 1990), *30*
Agreement to Establish Multilateral Fund (Copenhagen, 1992), *30*
agriculture, *181*, 187, 326
   international meetings about, 15, 17, 18
   in LDCs, 261, 271–72, 274, 288
   and prospects for ecologically sustainable development, 288
   subsidies for, 246
   and tax policies, 246, 248
   See also food

345

INDEX

air pollution, 4, 27–29, 106–7, *110*, 138, 169*n*42, 190, 202. *See also* acid rain; atmosphere; greenhouse gases; ozone layer; *specific agreement*; stratospheric ozone layer
Air Quality Agreement (U.S.-Canada, 1991), 27–29
Albright, Madeleine K., 195
Alliance for Responsible CFC Policy, 33
Alliance of Small Island States, 43–44, 45
Altiplano (Bolivia), 291–92
Amazonia, 291
American Petroleum Institute, 42
Anasazi people, 133–34
Anderson, Terry L., 167*n*19
Antarctic, 15, *22*, *60*, 263
Arctic, *181*
Asian Development Bank, 203
assessment
  of assistance, 275
  of climate change agreements, 45–48
  of environmental risk, 308
  of forests, 17, 207
  of international agreements, 72–74, 197–99
  of international organizations, 192–96
  international organizations role in, 198
  of NGOs, 208–9
  See also evaluation
assistance, 93, 163, 242, 262, 264, 288, 298, 326
  amount of, 270, *285*
  assessment of, 275
  and compliance, 200, 202
  and debt, 280–81, 285–86
  for funding implementation of international agreements, 280–86
  and GNP, 284, *285*
  and goals for agreements, 93, 105, 113, 115
  and implementation of agreements, 204–5, 275–97
  from international organizations, 200–205, 279, 281, 284
  levels of, 282–86
  and microloans, 295–97
  by private sector, 163, 270, 284
  regional, 189, 203, 206, 279, 283
  role of state in, 163
  for women, 293–97
  See also economic incentives; incentives; subsidies; technology; *specific agency, bank, or organization*
atmosphere, 4, 192, 263
  international agreements about, 20, 23–48
  UN bodies responsible for, *181*
  See also acid rain; air pollution; climate change; greenhouse gases; nuclear testing; ozone layer; stratospheric ozone layer; *specific agreement*
Ayres, Edward, 336–37

Bank for European Reconstruction and Development (BERD), 189, 191
banks, 191, 241–42, 295–97. *See also* debt; *specific bank or type of bank*
Benedick, Richard Elliot, 34
Beni Biosphere Reserve (Bolivia), 281
Berry, Wendell, 165, 331–32
biological diversity, *182*, 185, 245, 326
  and assistance, 201, 202–3, 204, 281, 282
  and "Earth Summit," 17, 18, 59, 88
  and economic incentives, 156–57
  and global agenda, 263
  international agreements about, 22, 56–63, *60–62*
  in LDCs, 263, 272, 281, 282
  and regulation, 156–57
  See also specific agreement
birds, *21*, *60*
Bolivia, 281, 291–92
Border Environmental Cooperation Committee (BECC) (NAFTA), 228
Boundary Waters Treaty (1909), 19
Boutros-Ghali, Boutros, 194, 205–6
Boyan, A. Stephen, 326–27, 328
Brady, Nicholas, 280
Brazil, 44, 46–47
Brown, Lester, 274
Bruntland Commission, 297
Btu taxes, 150
Business Council for a Sustainable Energy Program, 44
business-government cooperation, 242

Cahn, Matthew, 325–26
Cairo Conference on Population and Development (1994), 67–71

346

Index

California, 144–45, 170*n*61
Canada, 19, 25, 26–29, 57–58, 142, 305
Capacity 21, 189
car emissions, 144–45, 245, 249
carbon dioxide, 2, 25, 56–57, 158, 159, 162, 261. *See also* global warming
carbon sinks, 39, 57, 58
carbon taxes, 159, 164, 246
Catholic Church, 68–70
Central Europe, 189, 191
chemicals, 18, 107, 196, 248, 272, 276. *See also* hazardous waste; ocean dumping
Chernobyl, 125, 126
China, 43, 44, 46–47, 64, 65, 158, 263–64, 276
chlorofluorocarbons (CFCs), 107, 162, 201, 311
   and acid rain, 29–30, 31, 32–36, 32, 78*n*36
   alternatives to, 235, 281
   and global agenda, 263–64
   and global warming, 46, 88
   in LDCs, 263–64, 276, 281
   and technology transfers, 276
Clean Air Act (U.S.), 27, 28, 33, 76–77*n*28, 143–44, 154, 227
climate change, 107, 192, 286
   and acid rain, 25, 29
   and assistance, 201, 202–3, 281, 282
   costs and benefits of future, 162–63
   and deforestation, 56–57, 58
   international agreements about, 36–48, 38–39, 88, 107, 116, 158–59
   and LDCs, 9, 116, 258, 259, 271, 281, 282
   See also global warming; greenhouse gases; ozone layer; stratospheric ozone layer; *specific agreement or organization*
Climate Institute, 37
Climate Summit, 44, 45
coal taxes, 159
Cobb, John B., 249, 327
command-and-control regulation approach, 134–38, 140, 141, 151, 154, 245
commercial banks, 201, 270, 280, 295
Commission for Baltic Marine Environment Protection, 189
Commission on Human Rights (UN), 179

Commission on Narcotic Drugs (UN), 179
Commission for Social Development (UN), 179
Commission on the Status of Women (UN), 179
Commission on Sustainable Development (UN), 17, 187–88, 207
Committee on Challenges of Modern Societies (NATO), 177
Committee on Environmental Cooperation (CEC) (NAFTA), 228
Committee on International Development Institutions on the Environment, 192
Committee on Trade and the Environment (CTE) (GATT), 223
Commoner, Barry, 233–35, 237, 238–39
community, 295, 327. *See also* local efforts
community banks, 295–97
compliance, 9, 23, 73, 113, 140, 198
   and assistance to LCDs, 200, 202
   content of systems of, 119–20, 122–23
   costs of, 141, 150, 154
   and effectiveness, 123
   encouragement of, 93, 106, 123–27
   and goals for agreements, 89, 93, 95, 104, 105, 106, 116–27
   and implementation, 93, 116–27
   incentives for, 23, 89, 95, 104, 105, 106, 116–23, 198
   and LDCs, 200, 202, 268–69
   levels of complexity of, 116–18
   monitoring of, 93, 116, 119–20, 137
   and moral force, 123–24
   and noncompliance response mechanisms, 119–22, 123, 137
   and policy-making process, 91, 93, 116–27
   and requirements for international agreements, 118–23
   verification of, 93, 121, 122
   See also data collection; regulation; reporting; sanctions
Conference on the Changing Atmosphere (Toronto, 1990), 37
Conference on the Environment and Development (UNCED). *See* "Earth Summit"

INDEX

Conference on Environmentally Sustainable Development (World Bank, 1996), 163, 284
Conference on Global Development Cooperation (UN, 1993), 205–6
Conference on the Human Environment (Stockholm, 1972), 14–16, 125, 207
Conservation International, 281
Consultation Group on International Agricultural Research (CGIAR) (UN), *181*, *182*
consumption, 17, 108, 261–63, 326, 330, 336
Convention on Assistance in the Case of a Nuclear Accident or Radiological Emergency (1986), *21*
Convention on the Ban of the Import into Africa and the Control of Transboundary Movements and Management of Hazardous Wastes within Africa (Bamako, 1991), *21*, 71–72
Convention on Biological Diversity (Rio de Janeiro, 1992), *22*, 58–59, 62–63, *62*, 200–201, *202*
Convention on Civil Liability for Oil Pollution Damage (CLC) (1969), *49*
Convention for Co-operation in the Protection and Development of the Marine and Coastal Environment of the West and Central African Region (1981), *52*
Convention Concerning the Equitable Distribution of the Waters of the Rio Grande for Irrigation (1906), 19
Convention Concerning the Protection of the World Cultural and Natural Heritage (Paris, 1972), *22*, *61*
Convention on the Conservation of Antarctic Marine Living Resources (Canberra, 1980), *22*, *60*
Convention on the Conservation of Migratory Species of Wild Animals (CMS) (Bonn, 1979), *22*, *61*
Convention on the Continental Shelf (1958), *20*, *49*
Convention on the Control of Transboundary Movements of Hazardous Wastes and their Disposal (Basel, 1989), *21*

Convention on the Early Notification of a Nuclear Accident (1986), *21*
Convention on Environmental Impact Assessment in a Transboundary Context (Espoo, 1991), 71
Convention on the Establishment of an International Fund for Compensation for Oil Pollution Damage (Brussels, 1971), *49*
Convention of Fishing and Conservation of the Living Respources of the High Seas (1958), *20*, *49*
Convention on International Trade in Endangered Species of Wild Fauna and Flora (CITES) (Washington, 1973), *22*, *61*
Convention on Long-Range Transboundary Air Pollution (LRTAP) (1979), *20*, 25–26, *25*
Convention on Nature Protection and Wild Life Preservation in the Western Hemisphere (Washington, 1940), *21*, *60*
Convention on Oil Pollution Preparedness, Response, and Cooperation (OPRC) (London, 1990), *21*, 49–50
Convention for the Preservation and Protection of Fur Seals (1911), *21*, *60*
Convention for the Prevention of Marine Pollution by Dumping from Ships and Aircraft (Oslo, 1972), *50*, *51*
Convention on the Prevention of Marine Pollution by Dumping of Wastes and Other Matter (London, 1972), *20*, *50*, 72–73
Convention for the Prevention of Marine Pollution from Land-Based Sources (Paris, 1974), *21*, *51*
Convention for the Prevention of Pollution from Ships and the Protocol of 1978 Relating Thereto with Annexes (MARPOL) (1973, 1978), *21*, *50*, 72, 201–2
Convention for the Prevention of Pollution of the Sea by Oil (1954), *20*, *22*, *49*
Convention on the Protecting of the Black Sea against Pollution (1992), *52*

Index

Convention for the Protection of Birds (Paris, 1950), 21, 60
Convention for the Protection and Development of the Marine Environment of the Wider Caribbean Region (1983), 51, 52
Convention for the Protection, Management, and Development of the Marine and Coastal Environment of the Eastern African Region (1985), 52
Convention for the Protection of the Marine Environment of the Baltic Sea Area (Helsinki, 1972, 1974, 1992), 21, 51
Convention for the Protection of the Marine Environment and Coastal Area of the South East Pacific (1981), 52
Convention for the Protection of the Marine Environment of the North East Atlantic (OSPAR) (1992), 51
Convention for the Protection of the Mediterranean Sea Against Pollution (1976), 21, 52
Convention for the Protection of the Natural Resources and Environment of the South Pacific Region (1986), 52
Convention for the Protection of the Ozone Layer (Vienna, 1985), 20, 30, 31–32
Convention on Protection of the Rhine Against Chemical Pollution (1976), 21
Convention on the Regulation of Antarctic Marine Living Resources (1988), 60
Convention for the Regulation of Whaling (Washington, 1946), 20, 21, 52, 60
Convention Relating to Intervention on the High Seas in Cases of Oil Pollution Casualties (Brussels, 1969), 49
Convention on the Territorial Sea and the Contiguous Zone (1958), 20, 49
Convention to Combat Desertification (CCD) (Paris, 1994), 62
Convention on Wetlands of International Importance especially as Waterfowl Habitat (Ramsar, 1971), 22, 61
conventions, 20, 22–23. *See also specific convention*
corporations, 233–39, 241. *See also* multinational corporations (MNCs)

Daly, Herman E., 249, 324, 327
data analysis, 120, 121
data collection, 15, 17, 23, 45, 93, 120, 121, 122, 206–7, 251
debt
  and assistance, 280–81, 285–86
  and future costs and benefits, 249
  GNP ratio, 270
  of LDCs, 115, 259–60, 270–71, 272, 273, 280–81, 285–86, 294, 320
  swapping of, 201, 280–81
  of Third World, 281
decision making, 6, 164–65, 337
deforestation, 6, 18, 46, 56–63, 115, 158, 162, 186–87, 282, 326. *See also* forests; *specific agreement*
dependency theory, 287
deposit-return system, 146
desertification, 3, 15, 56–63, 62, 282. *See also specific organization or agreement*
developing countries. *See* less-developed countries; Third World
development
  alternatives to Western model of, 289–93
  comprehensive, 289–90
  ecologically sustainable, 286–93, 297–98
  and incrementalism, 298
  integration of environmental protection and, 88, 176, 188, 298
  and levels of aid, 283–84
  priorities for, 286
  social, 15, 108, 202
  and technology, 286–87, 289, 290, 293, 294, 297–98
  Western model for, 286–87
  women's role in, 288, 292, 293–97
  See also policies; *specific type of development or funding source*
development banks, 189, 203, 283, 296–97
dispute resolution, 23, 93, 105, 124–25, 197. *See also specific agreement or organization*
dolphins, 53, 224–25, 226, 253n33
Dubos, René, 15

INDEX

"due diligence," 124
Durning, Alan, 271

"Earth Charter" (1992), 16–19, 88
Earth Council, 192
"Earth Summit" (UNCED) (Rio de Janeiro, 1992), 5, 88, 207, 305
  agenda of, 16–19, 87, 88
  agreements resulting from, 16–18
  and assistance, 200, 205, 281, 284
  challenges in achieving goals for, 87–88
  and indigenous people, 332–33
  and intergenerational rights, 336
  and LDCs, 18, 262–63, 278–79, 281
  and national sovereignty, 16, 111
  and new environmental paradigm, 332–33, 336
  organizations spawned as result of, 19, 187, 188
  purpose of, 16
  report of, 105–6, 197–99, 278–79
  study of international agreements by, 105–6, 197–99
  as success or failure, 16, 18, 193
  See also Agenda 21 (1992); "Earth Charter" (1992); *specific topic*
Earth Watch (UN), *185*, 207
Easterbrook, Gregg, 314–15
Eastern Europe, 74, 112, 189, 191, 192, 234, 241, 282
eco-nomics, 239–51, 329–30
ecofeminism, 327–29
Ecological Studies Institute, 192
economic commissions, *20*, 179, 189, 190
economic incentives, 97, 217, 222
  advantages/disadvantages of, 155–57
  in Canada, 142
  efficiency of, 139–40
  and enforcement, 154–55, 160
  and EPA, 143–44, 149, 154, 168n30
  and greenhouse gases, 157–64
  and high prices, 248–49
  and implementation, 119, 139–43, 157–65
  market-based, 139, 140, 141, 143–48, 149–57, 164, 167n19, 217
  and moral force, 139, 155, 157
  and public support, 156
  and self-interest, 139
  and technology, 136, 142, 147, 151, 154, 158, 161, 163
  and true costs, 150–51, 164
  See also emissions: fees/taxes for; emissions: trading
economic issues, 3, 15, 66, 74, 126, *185*
  environment policies as disruptive for, 218, 240
  environmentalism as barrier to, 14–15
  and free market economy, 219–20
  and goals for agreements, 97, 104–5, 108, 115, 116, 119, 126, 127
  and growth as limitless, 222, 249, 315
  health versus, 218, 220
  and high prices, 248–49
  integration of environmental and, 108, 164, 244
  and national sovereignty, 225–26, 236
  and new environmental paradigm, 324, 325–26
  and new international economic order, 115
  and politics, 215–56
  and regulation, 215–56
  and technology, 219, 220, 232–39
  and true costs, 150–51, 164, 219, 221, 243, 245, 247, 248
  See also assistance; eco-nomics; economic incentives; less-developed countries; more-developed countries; national wealth; poverty; subsidies; tax policies; trade
Economic and Social Council (UN), 178, 187–88
Ecosystem Conservation Group (ECG) (UN), *182*, 192
elections of 1994 (U.S.), 138
emissions
  car, 144–45, 245, 249
  fees/taxes for, 140, 141–42, 144, 151, 156, 160–61, 164, 167n23, 246, 247
  sulfur, *20*, 28, 76–77n28, 143–44
  and technology, 232–33
  trading, 28–29, 77n28, 141, 142–45, 150, 151–56, 163, 167n19, 168n30
  See also type of emission
endangered species, 15, 18, *22*, *61*, 245. *See also* extinction

Index

energy, *184*, 261, 298, 311
  and assistance, 202
  in LDCs, 202, 275, 276–77
  and tax policies, 247–48
  and technology transfers, 275, 276–77
  See also type of energy
enforcement, 23, 73, 137, 160, 199
  and economic incentives, 154–55, 160
  and economic versus environmental policies, 240, 245
  of emissions trading, 154–55
  judicial, 135, 336
  and tax policies, 245
  UN as mechanism for, 193
  See also compliance; incentives; sanctions; *specific agreement*
Enterprise, 54
enters into force, 22, 197
Environment Programme (Global Environmental Monitoring System) (UN), *177*
environmental damage/threat
  distribution of, 136, 138, 155
  and future costs and benefits, 248–51
  in LDCs, 271–72, 274–75
  questions about major issues in, 6–7
  responsibility for, 297–98, 320, *321, 322*
  reversibility/irreversibility of, 9, 307, 314–15
  six critical issues concerning, 4–5
  See also problems
environmental impact, 71, 185, 329
Environmental Law Center, *177*
Environmental Law Institute, 192
Environmental Liaison Center, *177*
environmental paradigm, new, 324–37
environmental protection
  and future costs and benefits, 248–51
  and harmonious living with environment, 215–17
  and interdependence of life, 313–14
  as luxury, 14–15
  as moral issue, 314
  optimist-pessimist debate about, 314–16
  and precautionary principle, 311, 316
  priorities for, 18, 312, 313, 320
  public support for, 316, *317, 318, 319,* 320, *321, 322, 323*
  responsibility for, 335
  ripple effects of, 9–10
  and road metaphor, 336–37
  six critical issues concerning, 4–5
  six questions central to, 7–8
  and stewardship ethic, 314
  and unlimited growth, 222, 249, 315
  and values, 307, 313, 315
  See also specific topic
Environmental Protection Agency, U.S. (EPA), 117, 137, 312
  and air quality standards, 27, 169n42
  complaints about, 135, 136
  and economic incentives, 143–44, 149, 154, 168n30
  and emissions trading/allowances, 77n28, 143–44, 154, 168n30
  and hazardous waste, 138, 147–48
  review of consequences of foreign aid projects by, 277
  and technology, 137, 277
environmental risk, 9, 306–16, 326
Ereira, Alan, 1
European Community (EC), 33, *177*, 189, 190
European Council, 190
European Court of Justice, 190
European Environmental Bureau (Brussels), *177*
European Investment Bank, 189, 191
European Parliament, 190
European Union (EU), 225, 228, 232
evaluation
  and assistance to LCDs, 204
  and goals for agreements, 74, 89, 91, 93–95, 106, 123–27
  and implementation, 94–95, 123–27
  and policy-making process, 91, 93–94
  promoting, 123–27
  See also assessment
exactions, 148
extinction, 2, 5, 6, 57, 261. *See also* endangered species
Exxon Valdez oil spill, 233

fees
  emissions, 140, 141–42, 144, 151, 156, 160–61, 164, 167n23, 246
  pollution, 141–42, 156
  water pollution, 167n23
First Conference of the Parties to the Rio Convention (Berlin, 1995), 43–45
fishing, 3, 4, *20*, 245. *See also* marine resources; tuna

INDEX

food, 3, 4, 15, 187, 264, 326
   and development, 294
   in LDCs, 267, 269, 274, 288, 289, 294
   and population, 65, 66
   and prospects for ecologically sustainable development, 288, 289
   See also agriculture; *specific organization*
Food and Agriculture Organization (FAO) (UN), 22, *61*, *177*, 180, *181*, *182*, *183*, *184*, 203, 207, 269
forests, 17, *61–62*, 88, *182*, 207
   and acid rain, 24, 25
   boreal, 57
   in Canada, 57–58
   and carbon dioxide, 56–57, 261
   and distribution of environmental problems, 115–16
   and global agenda, 263
   and global warming, 39, 42
   in LDCs, 261, 263, 271–72
   old-growth, 261
   and tax policies, 245, 247
   tropical rain, 4–5, 56, 57, 115–16, 261, 263
   warnings about, 3, 4–5
   See also deforestation; *specific agreement*
fossil fuels, 25, 42, 44, 107, 115, 235, 246, 247, 261, 290, 297. *See also* greenhouse gases
Fowler, Robert Booth, 330–31
Framework Convention on Climate Change (Rio de Janiero, 1992), *20*, *38–39*, 38–43, 202
France, *30–31*, 164
fresh water, *183*, 282. *See also* water resources
Friends of the Earth International, 192
funding. *See* assistance; *specific organization or agreement*

Gaia (Greek goddess), 327–28
General Accounting Office, U.S., 136, 137, 138
General Agreement on Tariffs and Trade (GATT) (UN), 180, *185*, 197, 223–27, 229, 230, 231, 232, 253n33, 263
generations. *See* intergenerational rights
Germany, 167n23, 240

global agenda, 263–64
Global Atmosphere Watch (WMO), 207
Global Change Action Plan (U.S.), 41–42
Global Environmental Facility (GEF), *39*, 41, 62, 201–5, 281–82
Global Ozone Trends Panel, 32–33
global warming, 2–3, 6, *25*, 116, 220
   and assistance, 201, 204
   costs of preventing, 162–63
   and "Earth Summit," 38, *38*, 40, 87–88
   and goals for agreements, 87–88, 107, 116
   international agreements about, 36–48
   and LDCs, 40, 41, 42–43, 44, 45–46, 201, 204, 261
   and MDCs, 39, 40, 44, 45–46
   and technology, 158, 161, 163, 235
   See also climate change; greenhouse gases
goals for agreements
   and assessment/evaluation, 74, 89, 91, 93–95, 106, 123–27
   and assistance, 105, 113, 115
   and compliance, 89, 93, 95, 104, 105, 106, 116–27
   and comprehensiveness, 97–98, 108
   and distribution of environmental problems, 115–16
   and domestic policies, 91, 104, 114, 117–18
   and economic issues, 97, 104–5, 108, 115, 116, 119, 126, 127
   and implementation, 89–128
   and incentives, 89, 95, 97, 104, 105, 106, 112–23, 124, 126
   and incrementalism, 89, 91, 96, 97–99, 128
   and integration of environmental and other policies, 88, 89, 104, 105, 106–8
   and LDCs, 113, 114–16
   and moral force, 123–24
   and national sovereignty, 95, 111–12, 126, 127
   and participation, 89, 105, 109–16
   and policy-making process, 89–103
   and politics, 88–89, 95, 96–97, 102–3, 114, 117–18, 128
   and ratification, 90, 92, 97
   as realistic, 98–99

Index

and regulation, 114, 118–23
and self-interest, 112, 113, 122, 126, 127
and social goals, 108
and technology, 100–101, 105, 107, 113, 115, 121
See also specific agreement
goals, environmental. *See* goals for agreements; *specific agreement, organization, or topic*
Gore, Al, 329–30
governments. *See* national governments
grant funding, 281–82
grassroots. *See* local efforts
green taxes, 243–48
greenhouse gases, 2, 9, 18, 73, 107
  and assistance, 202, 204, 282
  and CFCs, 162
  and costs and benefits of future climate change, 162–63
  costs of reducing, 158, 159
  and deforestation, 158, 162
  distribution of global, 158, 163
  and economic incentives, 157–64
  and enforcement, 160
  foreign direct investment in reduction of, 163
  international agreements about, 163–64
  in LDCs, 158, 159, 160, 202, 204, 275, 276, 282
  and MDCs, 158, 159
  policy options concerning, 161–62
  and politics, 161
  and population, 158
  sources of, 157
  and taxes, 159, 160, 245
  and technology, 158, 161, 163, 275, 276
  and Third World, 158
  top producers of, 46–47, *47*
  See also global warming

Habitat and Human Settlements Foundation (UN), *177*
Hardin, Garrett, 274
Hawken, Paul, 237, 239
hazardous waste, 4, 18, *21*, 71–73, 138, 145–48, *183*, 235–36. *See also* chemicals; radioactivity; *specific agreement, organization, or substance*

health, 18, 155, *184*, 218, 220, 294, 295, 307
Helsinki Commission (HELCOM), 189
human rights, 108, 126

IBRD. *See* World Bank
implementation of agreements
  and adjustment, 123–27
  and assessment/evaluation, 74, 94–95, 123–27
  and assistance, 204–5, 275–97
  centralized versus decentralized approach to, 164–65, 175–76, 210
  as challenge, 6, 88–89, 94–128
  complexity of, 116–18, 140
  and compliance, 93, 116–27
  definition of, 96
  and economic issues, 119, 139–43, 157–65, 241
  factors affecting success of, 101–3
  and goals for agreements, 89–128
  and incrementalism, 96, 97–99, 165
  and integration of environmental and other policies, 106–8, 218, 220
  and intergenerational rights, 334–35, 336
  and international organizations, 198, 204–5, 210–11
  and LDCs, 204–5, 258–97
  and limited governmental capacity, 99–100
  local efforts for, 164–65, 316
  monitoring of, 91
  of moving target, 96–97
  and national sovereignty, 111–12
  and new environmental paradigm, 334–35, 336
  organizations as support for, 102–3, 210
  and participation, 109–16
  pitfalls of, 94–103
  and policy-making process, 89–103
  and politics, 88–89, 95, 102–3
  prospects for effective, 127–28
  and public support, 164–65, 316
  questions for examining, 105–6
  and tax policies, 245
  and tractability of problem, 100–101
  UNCED recommendations about, 197
  See also specific agreement
Inca empire, 257–58

353

incentives, 7, 136, 242
  for compliance, 23, 89, 95, 104, 105, 106, 116–23, 198
  and conventional regulation, 135, 137
  and economic versus environmental policies, 240, 242, 243–48
  and goals for agreements, 89, 95, 97, 104, 105, 106, 112–23, 124, 126
  and implementation, 139–43, 157–64
  for LDCs, 114–16
  for participation, 112–16
  and reuse of products, 243
  tax policies as, 243–48
  See also economic incentives
incrementalism
  and goals for agreements, 89, 91, 96, 97–99, 128
  and implementation, 96, 97–99, 165
  inadequacy of, 7
  and new environmental paradigm, 324, 326, 329, 336
  and sustainable development, 298
India, 23–24, 44, 46–47, 75–76n17, 263–64, 278
indigenous people, 215–17, 261, 286–93, 305–6, 332–34
industrialized nations. *See* more-developed countries (MDCs)
information, 15, *184–85*, 295
  dissemination of, 93, 120, 121, 122, 198, 199, 206–8
  exchanges of, 119, 207, 210
  and goals for agreements, 93, 116, 120, 121, 122
  UN as coordinator of, 206–7
  See also data collection; reporting
inspections, 121, 122, 137. *See also* *specific agreement*
Institute for European Environmental Policy, 192
institution building, 195, 207–8, 275–80, 333
institutions. *See* institution building; international organizations; organizations; *specific organization*
insurance industry, 44
intellectual property, 180, 276
Inter-American Development Bank, *178*, 203
Inter-governmental Oceanographic Commission (IOC) (UN), *177*, *181*, *182*

Inter-governmental Panel on Climate Change (UNEP-WMO), 39, 42, 44–45, *181*, 186, 241
Inter-Secretariat Committee on Scientific Problems Relation to Oceanography (ICSPRO), *183*
Inter-Secretariat Group on Water Resources (UN), *183*
Interagency Working Group on Desertification (UN), *181*
interdependence of life, 313–14, 327–28
intergenerational rights, 262, 290–91, 334–36
International Agency for Research on Cancer (WHO), *177*
international agreements, 5, *110*
  assessment of, 72–74, 197–99
  and domestic policies, 91, 104, 114, 117–18, 196
  as failures, 73–74
  function of, 74
  ignoring of, 73
  LDC role in, 278
  negotiating, 6, 197
  and policy-making process, 89–95
  primary rules for, 122
  provisions for, 104, 197
  ratification of, 92
  requirements for, 118–23
  UNCED analysis of, 105–6, 197–99
  See also goals for agreements; implementation of agreements; *specific agreement or topic*
International Arctic Science Committee (IASC) (UN), *181*
International Association of Ecology, *177*
International Atomic Energy Agency (IAEA) (UN), *21*, 127, *177*, *178*, *183*, *184*
International Bank for Reconstruction and Development. *See* World Bank
International Board for Plant Genetic Resources (BPGR) (UN), *181*, *182*
International Civil Aviation Organization (ICAO) (UN), 180, *182*
International Commission on Global Governance, 192, 197
International Conference on Population and Development (ICPD), 67–71

Index

International Convention for the Conservation of Atlantic Tunas (ICCAT) (Rio de Janeiro, 1966), *60*
International Council for Environmental Law, *178*
International Council for the Exploration of the Sea, *177*
International Council of Scientific Unions, *177*, 192
International Court of Justice (IJC-World Court) (UN), 124–25, *177*, 333
International Development Association (IDA) (UN), 180, 186, 283
International Dolphin Conservation Act, 224–25
International Energy Agency, *177*
international environmental agency, public support for, 320, *323*
International Finance Corporation (IFC) (UN), 180, 186
International Fund for Agricultural Development (IFAD) (UN), 180, *184*
International Institute for Applied Systems Analysis, 103–4, 192
International Institute for Environment and Development, *178*
International Labor Organization (ILO) (UN), 180, *181*, *182*, *183*, *184*, *185*
international law, 123–27, 336. *See also* legal issues
International Maritime Organization (IMO) (UN), *20*, *21*, *49*, *50*, 180, *181*, *182*, 203
International Monetary Fund (IMF) (UN), 180, 186, 187, 197, 270–71, 283
International Nuclear Information System, *178*
International Ocean Institute, *178*
international organizations
  assessment of, 192–96
  assistance from, *178*, 200–205, 279
  as catalysts, 210–11
  centralized versus decentralized approach to, 175–76
  coordinating efforts by, 205–7
  credibility of, 194
  and enforcement, 199
  list of environmental, 176, *177*–78, *181*–*85*
  and local efforts, 208–10
  and national sovereignty, 199, 200, 210
  overview of, 176–92
  and policy-making process, 200
  and politics, 199–200
  rise of, 241
  role/functions of, 207–8, 210–11, 279
  shortcomings of, 196–200
  and technology, 200–205
  See also institution building; non-governmental organizations; *specific organization*
International Planned Parenthood Federation, 67
International Plant Protection Convention (1951), *21*, *60*
International Research and Training Institute for the Advancement of Women (INSTRAW), 178
International Scientific and Technical Advisory Panel, 202
International Seabed Authority, 55
International Social Science Council, *177*
International Telecommunication Union (ITU) (UN), 180, *182*
International Tropical Timber Agreement (ITTA) (Geneva, 1983), *61*–*62*
International Undertaking on Plant Genetic Resources (Rome, 1983), *22*, *61*
International Union for Conservation of Nature and Natural Resources, *178*
International Whaling Commission (IWC), *21*, *48*, *52*, *60*
Interparliamentary Conference on the Global Environment (1990), 37
investment resources, in LDCs 269–71

Japan, 71, 74, 240, 261
John Paul II, 68–70
judicial bodies, 135, 336. *See also* dispute resolution; enforcement; *specific body*

Kogi people, 1–2
Kuna Indians, 290–91

INDEX

land, 3, 57, 138, 147–48, *181*, 332, 333
LDCs. *See* less-developed countries
League of Nations, 54
Leal, Donald R., 167*n*19
legal issues, 197, 198, 279, 330, 336. *See also* international law
Leopold, Aldo, 327
less-developed countries (LDCs), 9, 14, 115, 189, 191
 and adaptive strategies, 257–58
 aid and assistance for, 18, 23, 24, 43, 200–205, 275–97
 brain drain from, 278
 compliance in, 200, 202, 268–69
 consumption in, 261–63
 debt of, 259–60, 270–71, *272, 273,* 280–81, 285–86, 294, 320
 and "Earth Summit," 16, 18, 262–63, 278–79, 281
 and economic versus environmental policies, 218–19, 221, 246
 environmental damage in, 271–72, 274–75
 and global agenda, 263–64
 and goals for agreements, 113, 114–16
 government and regulatory infrastructure in, 268–69
 and implementation, 204–5, 258–97
 incentives for, 114–16
 income distribution in, 265–66, 287–88
 investment resources in, 269–71
 MDCs' relationship with, 259, 260–61, 268–69, 270–71, 275–98, 326
 national sovereignty of, 263
 and new environmental paradigm, 324–25, 326, 330, 335
 population in, 64, 65, 66, 68, 71, 260, 261–63, 274, 282, 290, 295
 poverty in, 259, 264–68, 271–72, 274–75, 282
 problems of, 258–60
 and public support for environmental protection, 316, 320, *322*
 regulation in, 268–69, 279
 role in agreements of, 278
 sustainable development in, 286–93, 297–98
 and technology, 235, 269–70, 274, 286–87, 289, 290, 293, 294, 297–98

and UN, 194
women in, 266–68, 288, 293–97
See also Third World; *specific topic or natural resource*
liberalism, 326, 328
Lindblom, Charles, 100
local efforts, 5, 118, 164–65, 208–10, 243, 288, 289–93, 294–97. *See also* community
London Convention (1990). *See* Convention on Oil Pollution Preparedness, Response, and Cooperation (OPRC); Montréal Protocol
London Dumping Convention (1972), *20, 50,* 72–73
Los Angeles, California, 144–45
Lovelock, James, 327–28
Lowi, Theodore, 100

Man and the Biosphere Program (MAB) (UNESCO), *177, 181, 184*
Management Sciences for Health, 192
Marine Mammal Protection Act (U.S.), 224–25, 253*n*33
marine resources, 16, 18, *20–21, 22,* 48–56, *49–52. See also* dolphins; fishing; oceans; seals; tuna; water resources; whales; *specific agreement*
Mazmanian, Daniel, 100–101, 102–3
media, 91, 92, 93, 191
Merchant, Carolyn, 328
Mesa Verde National Park (Colorado), 133–34
Mexico, 19, 67, 225, 227–32
Mitchell, Ronald, 119–20, 122–23
Molina, Mario J., 29
monitoring, 45, 73, 197, 204, 207
 for compliance, 93, 116, 119–20, 137
 of emissions trading, 154–55
 and goals for agreements, 91, 93, 119–20, 122, 126–27
 of hazardous waste, 146, 148
 of implementation, 91
 by international organizations, 198, 207
 on-site, 121
 and policy-making process, 91, 93
 and provisions in international agreements, 23, 104
 self-, 137
 See also specific agreement

356

Index

Monsanto, 242
Montréal Protocol (Montreal and London, 1987, 1990), 5, 16, 20, 34, 39
  and global agenda, 263–64
  as moral force, 124
  multilateral fund of, 30, 34, 201, 205, 281, 282
  participation in, 30, 32, 35, 78n36
  purpose/goals of, 16, 30, 32–33, 78n36
moral force, 123–24, 139, 155, 157, 251, 314, 329–30, 337
more-developed countries (MDCs), 9, 158, 159, 191
  and distribution of environmental problems, 116
  and economic versus environmental policies, 219, 221
  and global warming, 39, 40, 44, 45–46
  and goals for agreements, 107, 108, 114, 115, 116
  and integration of environmental and other policies, 107, 108
  LDCs' relationship with, 259, 260–61, 268–69, 270–71, 275–98, 326
  and marine resources, 54, 55
  as overdeveloped, 262–63
  and population, 68, 70–71
  and tax policies, 243, 246
  and UN, 194, 196
  and World Bank, 186
  See also assistance; technology
multilateral funds, 30, 34, 201, 205, 281, 282
Multilateral Investment Guarantee Agency, 186
multinational corporations (MNCs), 117, 128, 191, 193, 220, 221, 241, 263, 279–80, 336
Myers, Norman, 326

National Academy of Sciences (U.S.), 42, 192
"national action plans," 17
National Ambient Air Quality Standards (U.S.), 27
National Association of Manufacturers, 33
national climate plans, 44

National Environmental Protection Act (U.S.), 233
national governments
  and assistance to LCDs, 204
  business cooperation with, 242
  limited capacity of, 99–100
  organizations of, 177
  See also national sovereignty
National Oceanic and Atmospheric Administration, 35
national sovereignty, 16, 263
  and economic versus environmental issues, 225–26, 236
  and goals for agreements, 95, 111–12, 126, 127
  and implementation, 111–12
  and international organizations, 199, 200, 210
  and trade, 225–26
national wealth, 250–51, 279–80
natural resources, 3, 15, 17, 326
  distribution of, 259
  as freely substitutable, 222, 249
  price of, 324
  unlimited, 222, 249, 315
  See also specific resource
network organizations, 210
New and Renewable Resources of Energy (NRSE) (UN), 184
newly industrialized nations, 221
nitrogen oxide, 20, 25, 77n28, 144–45, 155. *See also* global warming
nongovernmental organizations (NGOs), 15, 67, 73, 188, 193, 246, 280
  assessment of, 208–9
  and assistance, 203, 204, 284, 295–97
  and economic versus environmental policies, 241–42
  and goals for agreements, 91–93, 112–13, 122, 128
  importance of, 208–10
  and LDCs, 203, 204, 209
  and policy-making process, 91, 93
  and problem definition, 91
  rise of, 241–42
  role/functions of, 208–10
  selected environmental, 177
  and World Bank reorganization, 187
  See also specific organization

357

Nordic Investment Bank, 189
North American Agreement on Environmental Cooperation (NAAEC), 228
North American Free Trade Agreement (NAFTA), 227–32
North Atlantic Treaty Organization (NATO), *177*
North Korea, 72
North Pacific Fur Seal Commission, 48
North. *See* more-developed countries
Northeast States for Coordinated Air Use Management, 144, 155
notification of dangerous releases, 18, 23, 124
nuclear power, 15, *21*, 107, *110*, 164, 196. *See also* nuclear testing; *specific agreement*
Nuclear Test-Ban Treaty (1995–1997), 23–24, 75–76n17
nuclear testing, 15–16, *22*, 23–24. *See also specific agreement*

objectives, 88, 89, 106–8, 197
ocean dumping
  and Agenda 21, 17
  as coordination or collaboration problem, *110*
  international agreements about, 15, *20*, *50*, 53, 56, 72–73
  regional agreements about, *51*, *52*
oceans, 4, 6, *182–83*
  and assistance, 201–2, 204, 281
  and global agenda, 263
  regional, 185
  See also dolphins; fishing; marine resources; ocean dumping; oil/oil spills; seals; tuna; water resources; whales; *specific agreement*
Office for Research and Collection of Information (UN), 207
Office of the United Nations High Commissioner for Refugees (UNHCR), 179
oil/oil spills
  Exxon Valdez, 233
  and greenhouse gases, 159
  international agreements about, *20*, *22*, *49*, *50*, *110*, 119–20
  regional agreements about, *52*

  taxes, 159
  See also specific agreement
Only One Earth (Stockholm Conference, 1972), 15
Ophuls, William, 326–27, 331
Organisation for Economic Co-operation and Development (OCED), *177*, 190, 282
Organization of African Unity (OAU), 72
Organization of American States (OAS), *21*, *60*
organizations
  as catalysts, 210–11
  criticisms of environmental, 238–39
  management of, 330
  and new environmental paradigm, 330
  See also institution building; international organizations; non-governmental organizations; *specific organization*
Oslo Commission on Ocean Dumping (1972), *50*, *51*
Ostrom, Elinor, 109
Oxfam International, 270
ozone layer, 24, 36–37, 53, 106–7
  and assistance, 201, 204, 281, 282
  and emissions trading, 144
  and incentives, 144, 145
  international agreements about, 16, *20*, 73
  and LDCs, 201, 204, 258, 281, 282
  See also acid rain; global warming; greenhouse gases; stratospheric ozone layer; *specific agreement*
Ozone Projects Trust Fund (GEF), 205

paradigm, environmental, 324–37
Paris Club, 270–71
Partial Nuclear Test Ban Treaty (1963), 23
participation in agreements, 89, 105, 109–16, 197. *See also specific agreement*
"Partners in Population and Development: A South-South Initiative," 71
pesticides, 16, 53
policies, 14–15, 136, 210, 309
  as disruptive, 218, 240
  domestic, 91, 104, 114, 117–18, 196

358

Index

and environmental risk, 311–16
and goals for agreements, 91, 104, 114, 117–18
integration of environmental and other, 88, 89, 91, 96, 97–99, 105, 106–8, 128, 164, 200, 218, 220, 244
and politics, 8, 100, 312
and precautionary action, 8–9, 23, 311, 316
and road metaphor, 336–37
and scientific uncertainty, 306–16
and unlimited growth, 222, 249, 315
and values, 307, 313, 315
See also incrementalism; policy analysis; policy formulation; policy-making process
policy analysis, 94, 97
policy formulation, 90, 92, 97
policy-making process, 89–103, 116–27, 156, 200
politics, 6, 73, 161, 260
and conventional regulatory approaches, 135–36
and economic versus environmental issues, 215–56
and environmental risk, 312
and goals for agreements, 88–89, 95, 96–97, 102–3, 114, 117–18, 128
and implementation of agreements, 88–89, 95, 102–3
and international organizations, 199–200
and new environmental paradigm, 324, 326, 337
and policies, 8, 94, 100, 312
and public support, 316, 320
and regulation, 215–56
and technology, 232–39
See also national sovereignty; tax policies
pollution
as capitalist problem, 14
end of, 314–15
taxes, 141–42, 156, 160
See also type of pollution or specific agreement
pollution-control companies, 241
poor nations. *See* less-developed countries

population, 3, 5, 46, 158, *183*, 329
and abortion, 67, 68–70
and assistance, 262, 282
and consumption, 261–63
and contraceptives, 65–66, 69, 85n130
and economic issues, 66
international agreements/meetings about, 15, 16, 17, 22, 63–71
in LDCs, 64, 65, 66, 68, 71, 260, 261–63, 274, 282, 290, 295
and MDCs, 68, 70–71
and religious groups, 68–70
and technology, 66
and women's issues, 67–68, 70
See also food; *specific agreement*
Population Commission (UN), 179, *183*
Portney, Paul, 137
poverty, 3, 18, 219, 262
in LDCs, 14, 259, 264–68, 271–72, 274–75, 282
UN bodies responsible for, *183–84*
precautionary action, 8–9, 23, 311, 316
Pressman, Jeffrey, 96
"prior informed consent," 18
private sector, 163, 187, 204, 270, 284, 309. *See also* multinational corporations; nongovernmental organizations
problems
collaboration, 109–10
coordination, 109–10, 111
defining, 90, 91
distribution of environmental, 115–16
roots of environmental, 165
tractability of, 100–101
products, reuse of, 242–43, 329
Protocol Concerning the Control of Emissions of Nitrogen Oxides or Their Transboundary Fluxes (Sofia, 1988), *20*
Protocol Concerning the Control of Emissions of Volatile Organic Compounds or Their Transboundary Fluxes (Geneva, 1991), *20*
Protocol Concerning the Reduction of Sulphur Emissions or Their Transboundary Fluxes (Helsinki, 1985), *20*
Protocol on Environmental Protection to the Antarctic Treaty (1991), *22, 60*

INDEX

Protocol on Long-Term Financing of the Cooperative Programme for Monitoring and Evaluation of the Long-Range Transmission of Air Pollutants in Europe (Geneva, 1984), 25
Protocol Relating to Intervention on the High Seas in Cases of Pollution by Substances Other than Oil (1973), 50
Protocol on Substances that Deplete the Ozone Layer (1987), 16, 20, 30, 32–36, 39
Protocol to Freeze Nitrogen Oxide Emissions at 1987 Levels by 1994 (1988), 25
protocols, 20, 91, 197. *See also specific protocol*
public, 140–41, 156, 309
  and new environmental paradigm, 337
  and politics, 316, 320
  support for environmental protection by, 316, *317*, *318*, *319*, 320, *321*, *322*, *323*
  support for international environmental agency by, 320, *323*

radioactivity, 16, 17, *21*, 53
ratification, 90, 92, 97. *See also specific agreement*
Rawls, John, 335
recycling, 242–43, 329
regional assistance programs, 206, 279. *See also specific program or agreement*
Regional Clean Air Incentives Market (RECLAIM), 145, 170*n*61
Regional Convention for Co-operation on the Protection of the Marine Environment from Pollution (Kuwait, 1978), 52
Regional Convention for the Conservation of the Red Sea and Gulf of Aden Environment (1982), 52
regional development banks, 189, 203, 283
Regional Enviromental Center for Central and Eastern Europe, 189
regional organizations, 189–91, 195. *See also specific organizationn*
regional trade, 227–28, 232

regulation
  aim of, 219
  call for effective global, 198–99
  complaints/criticisms of, 134–38, 218–19
  conventional approach to, 134–38, 140, 141, 151, 154, 245
  costs and benefits of, 136, 137, 150–51, 164, 218–19, 220, 236, 239–41, 248–51
  and domestic policies versus goals for agreements, 117–18
  and economic issues, 215–56
  and global treaty requirements, 118–23
  and goals for agreements, 114, 117–23
  judicial enforcement of, 135, 168*n*30
  in LDCs, 268–69, 279
  and moral force, 139, 155, 157
  and policy-making process, 156
  and politics, 135–36, 215–56
  and prospects for eco-nomics, 239–51
  and public support, 140–41, 156
  and self-interest, 139
  and technology, 232–39
  and trade, 220–32
  See also economic incentives; emissions: trading; tax policies; *type of regulation*
religious groups/values, 68–70, 330–32
Repetto, Robert, 279
reporting, 104, 116, 120, 121, 122, 137
requirements for compliance, 23, 118–23
research, 73, 207. *See also* data analysis; data collection; policy analysis
resettlement, 15, 37, *177*, 179, *183*, 187
Resource Conservation and Recovery Act (RCRA), 138
Rio de Janeiro, Brazil. *See* "Earth Charter"; "Earth Summit"; First Conference of the Parties to the Rio Convention
Rio Declaration. *See* "Earth Charter"
road metaphor, 336–37
Rockefeller Brothers Fund, 192
Rowland, F. Sherwood, 29
Russell, Clifford, 145–46

Sabatier, Paul, 100–101, 102–3
sanctions, 23, 73, 105, 158
  and economic versus environmental policies, 223, 224, 240

Index

and goals for agreements, 93, 104–5, 116, 121–22, 123, 127
trade, 223, 224
Schumacher, E. F., 236
science, 6, 15, 268, 278, 306–16, 330. *See also* technology
Scientific Committee on Effects of Radiation (UN), *177*
seals, 21, *60*
self-interest, 112, 113, 122, 126, 127, 139, 314, 325–26, 336
Sierra Nevada de Santa Marta (Colombia), 1–2, 291
signatory, definition of, 21
Smil, Vaclav, 330
social development, 15, 108, 202
soil, 3, 4, 6, 16, 57–58, 263, 271–72
South Asia Cooperative Environment Program, *177*
South Coast Air Quality Management District (California), 144–45
South. *See* less-developed countries (LDCs); Third World
Soviet Union, 46, 158, 316
spiritual values, 330–32
State of the World Population (U.S. report), 64
stewardship ethic, 314
Stockholm conference (1972), 14–16, 125, 207
Stockholm Declaration on the Human Environment (1972), 15, 335
Stockholm Initiative on Global Security and Governance, 192, 196–97
"Strategic Environmental Initiative" (Gore), 329–30
stratospheric ozone layer, 9, 107, 282
international agreements about, 29–36, *30–31*, 185
and public policies, 308, 311
and scientific uncertainty, 311
warnings about, 2–3, 4
See also specific agreement
Strong, Maurice, 15, 192
subsidies, 57, 243, 245–46
sulfur emissions, *20*, *28*, 76–77n28, 143–44. *See also* acid rain
Summers, Lawrence, 324–25
Supreme Court, U.S., 168n30
sustained development
and assistance to LCDs, 201, 204

and "Earth Summit," 17, 19, 88
and international organizations, 198, 201, 204
and NGOs, 192
and World Bank/IMF projects, 187
See also Commission on Sustainable Development (UN)

Taiwan, 72
tax policies, 243–48, 255n84, 285, 329. *See also specific tax*
taxes. *See* tax policies; *type of tax*
technology, 15, 17, 66, 74, 134, 137, 242
assistance, 200–205, 242, 275–80
and corporations, 233–39
and debt swapping, 280
and development, 286–87, 289, 290, 293, 294, 297–98
development of new, 135, 136, 163
and economic incentives, 136, 142, 147, 151, 154, 158, 161, 163
and economic versus environmental policies, 219, 220, 232–39
and emissions, 232–33
and goals for agreements, 100–101, 105, 107, 113, 115, 121
and greenhouse gases, 158, 161, 163
and hazardous waste, 147
and industrialized nations, 234–35
and integration of environmental and other policies, 107
and international organizations, 200–205, 278–79
and LDCs, 235, 269–70, 274, 286–87, 289, 290, 293, 294, 297–98
and new environmental paradigm, 329, 330, 337
and politics, 232–39
and regional organizations, 189
and regulation, 232–39
and tractability of problems, 100–101
and traditional/indigenous knowledge, 306
transfer of, 17, 43, 59, 121, 200, 275–80, 297–98
UN bodies responsible for, *182*, *185*
See also science
Third World, 33, 46, 108, 158, 201, 263, 269, 279–80, 281. *See also* Agenda 21; less-developed countries

361

INDEX

Third World Network, 262–63, 336
Thornburg, Richard, 194
3M Company, 242
Tietenberg, Tom, 315
timetables in agreements, 23
Tolba, Mostafa, 34
toxic materials. *See* chemicals; hazardous waste
trade, 108, *185*, 187, 196, 250
  and comparative advantage, 222
  and economic versus environmental policies, 220–32
  embargos, 225, 227
  and LDCs, 221, 263, 286, 294
  and level of aid, 286
  and national sovereignty, 225–26
  and new enviromental paradigm, 336
  and pollution-control products, 241
  regional, 227–28, 232
  and regulation, 220–32
  sanctions, 223, 224
  and specialization, 222
  See also specific agreement or organization
"tragedy of the commons," 109–11
Transfer of Knowledge through Expatriate Nations (TOKTEN) (UNDP), 278
transportation, *184*
treaties, 20, 22–23. *See also specific treaty*
Treaty Banning Nuclear Weapons Tests in the Atmosphere, in Outer Space and Under Water (1963), 22
Tropical Forests Action Program (TFAP) (UN), *182*
true costs, 150–51, 164, 219, 221, 243, 245, 247, 248
tuna, 224–25, 232, 253*n*33

UNCED. *See* "Earth Summit"
UNCLOS. *See* United Nations Convention on the Law of the Sea
UNDP. *See* United Nations Development Program
UNEP. *See* United Nations Environment Program
United Nations
  agenda/functions of, 174–75, 194, 200
  Charter of, 173, 174, 178, 196, 335
  coordinating efforts by, 205–7
  creation of, 173
  criticisms of, 175–76, 193–99, 206–7
  and enforcement of agreements, 193
  environmental agencies and programs of, *177*
  estimates for development assistance by, 283
  examples of agencies and organizations of, 178–80
  General Assembly of, 24, 173, 178, 187, 188, 194, 195
  as implementation force, 210
  importance of, 176
  and LDCs, 194
  and MDCs, 194, 196
  membership in, 173
  organization of, 173–76, 178
  overview of, 178–89
  and policy-making process, 91, 92
  problems among agencies of, 176
  reform of, 194–99, 206–7
  Secretary-General of, 178, 188, 196
  Security Council of, 178, 196
  special observances of, 174–75
  strengths of, 193
  See also specific agency, agreement, commission, conference, program, or affiliated organization
United Nations Center for Science and Technology Development (UNCSTD), *185*
United Nations Center on Transnational Corporations (UNCTC), *185*
United Nations Children's Fund (UNICEF), 179, *181*, *184*, *185*, 282, 293–94
United Nations Commission for Human Settlements (UNCHS), *177*, 179, *183*
United Nations Conference on Trade and Development (UNCTAD), *177*, 179, *182*, *185*, 189
United Nations Convention on the Law of the Sea (UNCLOS), 5, *21*, *50–51*, 54–56, 72, 193, 202, 226
United Nations Development Program (UNDP), *177*, 179, 286
  and assistance, 201, 202, 203
  Human Development Report of, 264–66
  proposed reform of, 197–98, 278

Index

responsibilities/focus of, *181*, *182*, *183*, *184*, *185*, 188–89
United Nations Educational, Scientific, and Cultural Organization (UNESCO), 22, 177, 180, *181*, *182*, *183*, *184*, *185*
United Nations Environment Program (UNEP), 15, 179
  and assistance, 201, 202, 203, 205
  overview of, 184–86
  proposed reform of, 197–98, 278
  Regional Seas Programs, 19, 22, 51–52
  responsibilities/focus of, *181*, *182*, *183*, *184*, *185*, 188, 207
  See also specific agreement
United Nations Fund for Population Activities (UNFPA), 46, 67, 85n130, 158, 177, 179, *183*, *185*, 282
United Nations Industrial Development Organization (UNIDO), 177, 180, *181*, *182*, *183*, *185*
United Nations Institute for Training and Research (UNITAR), 177, 179
United Nations Office for Ocean Affairs and the Law of the Sea (OALOS), 21, *183*
United Nations Relief Works and Agency for Palestine Refugees in the Near East (UNRWA), 178
United Nations Statistical Commission, 179, 207
United Nations Sudano-Sahelian Office and UN Statistical Office (UNSO), *181*, *184*
United Nations University (UNU), 179
United States
  Canadian agreements with, 25, 26–29
  consumption in, 261
  environmental problems in, 261
  Mexican agreements with, 19, 227–32
  role in environmental protection of, 9
  See also specific topic
Universal Declaration of Human Rights, 335
Universal Declaration on the Rights of Indigenous Peoples, 332
Universal Postal Union (UPU) (UN), 180

values, 307, 313, 315, 326, 330–32, 333
verification, 93, 121, 122
Vice President's National Performance Review (Gore, 1993), 137

Wallerstein, Immanuel, 326
Ward, Barbara, 15
"Warning to Humanity" (1993), 3–5
wars, 269, 284, 326
water resources, 3, 4, 25, *110*, 167n23, 207, 258. *See also* fresh water; water; marine resources; oceans
wealth, national, 250–51, 279–80
wealthy nations. *See* more-developed countries
Weeden Foundation, 281
Weiss, Edith Brown, 333–34
wetlands, 22, 53, *61*
whales, 16, *20*, *21*, 48, *52*, *60*. *See also* International Whaling Commission; marine resources
Wildavsky, Aaron, 96, 310
wildlife, *21*, 22, 57, *60*, *61*, 178. *See also* deforestation
Wilson, E. O., 56
Wolfensohn, James D., 187
women
  and Agenda 21, 17
  assistance to, 293–97
  and banks, 295–97
  and ecologically sustainable development, 292
  international meetings about, 15
  in LDCs, 266–68, 288, 293–97
  and new environmental paradigm, 327–29
  and population, 67–68, 70
  role in development by, 288, 293–97
Women's Initiative for Self-Employment, 297
Woodhouse, Edward, 100
World Bank (International Bank for Reconstruction and Development)
  conference on sustainable development (1996) of, 163
  creation of, 186
  criticisms of, 186–87, 197
  and debt of LDCs, 270–71
  and future costs and benefits, 250
  and GEF, 201, 202, 203

INDEX

World Bank (*continued*)
  internal study of, 187
  as international funding source, *178*,
    180, 186–87, 203, 205, 281–82
  and level of aid, 283, 284
  MDC domination of, 186
  and new environmental paradigm,
    324–25
  overview of, 186–87
  projects of, *39*, 40, 186–87
  and regional organizations, 189, 191
  and technology transfers, 276
World Conservation Union (IUCN),
  22,191–92
World Court. *See* International Court of
  Justice (IJC-World Court) (UN)
World Environment Center (New York),
  178, 192
World Food Council (WFC), 179
World Food Program (WFP) (UN-FAO),
  179, *181*, *182*, *184*

World Health Organization (WHO)
  (UN), *39*, *177*, 180, *182*, *183*,
  184, 203, 207
World Heritage Convention, *22*, *61*
World Intellectual Property
  Organization (WIPO) (UN), 180
World Meteorological Organization
  (WMO) (UN), *177*, 180, *181*,
  182, *183*, *184*, 186, 207
World Resources Institute, 195–96,
  282–83
World Summit on Global Governance,
  197
World Trade Organization (WTO),
  226–27
World Wildlife Fund, *178*
World Wood Program, 283
WorldWatch Institute, 2–3, 283

Yunus, Muhammad, 295–96